Space Education & Strategic Applications

Also from Westphalia Press
westphaliapress.org

SE:SA

SPACE EDUCATION & STRATEGIC APPLICATIONS

VOLUME 3, NUMBER 2 • WINTER 2022

Dr. Kristen Miller and Dr. Gary Deel, *Co-Editors-in-Chief*

Westphalia Press
An imprint of Policy Studies Organization

SPACE EDUCATION & STRATEGIC APPLICATIONS
VOL. 3, NO. 2 • WINTER 2022

All Rights Reserved © 2023 by Policy Studies Organization

Westphalia Press
An imprint of Policy Studies Organization
1367 Connecticut Avenue NW
Washington, D.C. 20036
info@ipsonet.org

ISBN: 978-1-63723-945-2

Cover and interior design by Jeffrey Barnes
jbarnesbook.design

Daniel Gutierrez-Sandoval, Executive Director
PSO and Westphalia Press

Updated material and comments on this edition
can be found at the Westphalia Press website:
www.westphaliapress.org

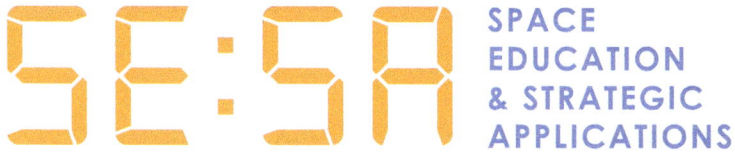

SE:SA

SPACE EDUCATION & STRATEGIC APPLICATIONS

SPACE EDUCATION AND STRATEGIC APPLICATIONS JOURNAL

VOL. 3, NO. 2 / WINTER 2022

© 2023 Policy Studies Organization

Letter from the Editors – January 2023

L ast year marked the 50th anniversary of the Apollo 17 mission—the last time that mankind set foot on the Moon. This year is the 50th anniversary of Skylab, the first American space laboratory and precursor to the International Space Station. These landmarks in the history of space exploration, and the years of research in lower Earth orbit (LEO) which followed as a result, prepared the way for the beginning of a new era in manned space exploration with the successful launch and return of the Artemis I mission on November 16, 2022. The Artemis missions will lead the way for humanity's return to the Moon, this time to stay, and will also be the first step toward human exploration and perhaps one day settlement of Mars. As the culmination of 50 years of technological development and research, Artemis represents not only a return to the Moon, but also a step forward for humanity with a realization of—and a firm commitment to—the importance and benefits of diversity.

The 21st century has also been a new era in commercial spaceflight. Commercial companies are providing opportunities for private citizens to experience microgravity through both suborbital and orbital experiences. More than 20 space tourists have now experienced microgravity through five suborbital Blue Origin flights (Blue Origin). In 2021, the first ever full non-professional crew flew aboard SpaceX's Crew Dragon to successfully complete the Inspiration 4 mission which spent three days in orbital space (Inspiration 4, 2022). Other companies, such as Virgin Galactic, are set to begin regular space tourism flights in the coming year as well.

International participation in space exploration has also increased; there are currently 86 countries with space capabilities worldwide. Of these, eleven countries have the ability to send payloads to space and three (the U.S., Russia, and China) have successfully sent human beings to outer space. More than 260 people from 20 different countries have lived aboard the International Space Station (ISS). Lower Earth Orbits are truly an international domain, and farther destinations such as the Moon and Mars will soon follow suit.

This new era of space exploration brings with it an array of new challenges and the need for new policies, technologies, and medical remediations to govern the use of space resources, protect astronauts from the dangers of the space environment and preserve both scientific and environmental agendas on other worlds as well as our own. In addition, it is becoming increasingly important to develop robust security measures to protect national assets in space. Current space policies and treaties will need to evolve to address these challenges, and international agreement on space policies is vital.

In this exciting and rapidly evolving space environment, the editors are pleased to present the third issue of the Space Education and Strategic Applica-

doi: 10.18278/sesa.4.1.1

tions (SESA) Journal, together with the American Public University System and the Policy Studies Organization. The articles presented in this issue directly speak to the challenges and opportunities of the current space environment. Dr. Chapman's analysis of the evolving domain and policies of the U.S. Space Force speaks to its essential role in securing national resources in the increasingly complex international arena of space, particularly in lower earth orbits where satellites and sensors vital to national security reside (Chapman, 2022). Duke (2022) also addressed issues of national security in space in near Earth, lunar, and Martian environments. The author underscores the pivotal role space plays in national defense, the dangers of space warfare and anti-satellite weapons, and the implications of exploitation of space resources. Taylor Nichols' article looks forward to the potential for human exploration of Mars and presents an analysis of the effects of global environment conditions on radio propagation. This article examines the potential to apply GIS modeling to Mars to better understand the communications challenges that future missions will face (Nichols, 2022).

We are grateful to our SESA authors and our SESA readers for their participation and support of this publication. We look forward to continuing a robust future of contributions to the space research community with this and future issues.

Kristen Miller & Gary Deel
Co-editors in chief, *SESA Journal*

References

Blue Origin. (2022). The Latest from Blue. Retrieved December 5, 2022, from https://www.blueorigin.com/news/

Chapman, B. T. (2022). Biden Administration U.S. Space Force policy literature. *Space Education & Strategic Applications,* December. https://doi.org/10.18278/sesa. 4.1.4

Duke, J. E. (2022). Seizing the stars: Resources, expansion, and counterspace contingencies across the space domain. *Space Education & Strategic Applications,* November. https://doi.org/10.18278/sesa.4.1.1.

Inspiration 4. (2022). The first all-civilian crew to orbit. Retrieved December 5, 2022, from https://inspiration4.com/

Nichols, T. (2022). Toward an integrated GIS Model of the Martian communications environment. *Space Education & Strategic Applications,* November. https://doi.org/10.18278/sesa.4.1.2.

Space Education and Strategic Applications Journal • Vol. 3, No. 2 • Winter 2022

Toward An Integrated GIS Model of the Martian Communications Environment

Taylor Nichols

ABSTRACT

Given the essential role of communications within a space exploration mission, it is important to accurately plan communications systems around the environmental parameters they will meet with on mission location. Mars is the expected goal of many future missions, and as such requires an accurate model of radio propagation within its environment. Extant models with this goal tend to be highly localized. In this study, a global scale GIS model of the Martian communications environment is built using existing raster and vector data implemented in the QGIS program. Overlaying these properties created a map depicting favorable and unfavorable communication conditions across the Martian globe. The global scale and multilayered nature of this depiction can help mission planners understand and prepare for large scale environmental patterns affecting communications on Mars.

Keywords: communications, GIS, Mars, Martian environment

Hacia un modelo SIG integrado del entorno de comunicaciones marciano

RESUMEN

Dado el papel esencial de las comunicaciones dentro de una misión de exploración espacial, es importante planificar con precisión los sistemas de comunicaciones en torno a los parámetros ambientales con los que se encontrarán en el lugar de la misión. Marte es el objetivo esperado de muchas misiones futuras y, como tal, requiere un modelo preciso de propagación de radio dentro de su entorno. Los modelos existentes con este objetivo tienden a estar muy localizados. En este estudio, se construye un modelo SIG a escala global del entorno de comunicaciones marciano utilizando datos ráster y vectoriales existentes implementados en el programa QGIS. La superposición de estas propiedades creó un mapa que representa las condiciones de comunicación favorables y desfavorables en todo el

doi: 10.18278/sesa.4.1.2

globo marciano. La escala global y la naturaleza multicapa de esta representación pueden ayudar a los planificadores de misiones a comprender y prepararse para los patrones ambientales a gran escala que afectan las comunicaciones en Marte.

Palabras clave: comunicaciones, SIG, Marte, entorno marciano

迈向关于火星通信环境的GIS综合模型

摘要

鉴于通信在太空探索任务中的关键作用，对通信系统将在飞行任务位置遇到的环境参数进行准确的通信系统规划一事非常重要。火星是许多未来太空飞行任务的预期目标，因此需要在其环境中建立准确的无线电传播模型。以此为目标的现有模型往往是高度本地化的。本研究中，使用在QGIS程序中执行的现有栅格数据和矢量数据，构建了一个关于火星通信环境的全球地理信息系统（GIS）模型。通过叠加这些属性进行制图，描绘了火星全球范围内有利和不利的通信条件。这一描述的全球规模以及多层次性质能帮助太空飞行任务规划者理解一系列影响火星通信的大规模环境模式，并为之做好准备。

关键词：通信，地理信息系统，火星，火星环境

Introduction

With the eyes of the scientific community turned expectantly towards Mars as the destination of a majority of future missions, it seems increasingly relevant to find ways to improve the stability of communications within that environment in order to ensure mission success. The communications system is the only means of connecting the mission to Earth. Without the ability to communicate, a mission cannot be conducted in any way that matters, or often even in any way at all. The very life of the mission depends upon it. Because the Martian environment is different from that of Earth in almost every respect, radio waves (the established method of current space communications) will react and therefore propagate differently there than they do on Earth. The multiplicity of parameters involved beg synthesis, and for that purpose this study proposes to create an integrated model of the communication-affecting properties in the Martian environment with the aid of a Geographic Information System (GIS) program. A GIS program is capable of synthesizing a vast array of independent, spatially-defined data sets in order to identify patterns within the environment. This study intends to use GIS to overlay maps of different environmental properties on Mars in order to build a model of the Martian communication environment. Such a model provides useful insight into such mission planning efforts as identifying potential communications hazards, defining communications zones, and determining which propagation parameters are most affected for missions in different areas of Mars.

Background

Perhaps the most comprehensive work on Martian communications is the *Radio Wave Propagation Handbook for Communication on or Around Mars* by Kliore et al. (2002). It serves as an in-depth analysis of the potential Martian communications environment. Since its publication, however, more accurate and up-to-date environmental data has been obtained from missions like MAVEN or Mars Express, which has contributed to the creation of various new environmental models. These have been applied to communications modeling in a hitherto detailed but limited scope. Chukkala et al. (2004; 2005), Borah et al. (2005), Daga et al. (2006), and more recently Bonafini and Sacchi (2020a; 2020b; 2021) have all successfully experiment-

ed with using various terrestrial modeling software altered according to Martian parameters to define communication potential in particular Martian zones. While more ambitious global models do not tend to cover communication aspects of the environment, they have successfully used GIS platforms to analyze specific characteristics of the Martian terrain on a global scale (see Kuziakina et al., 2019; Kuzma et al., 2018). This study follows the method established by Kuziakina et al. (2019) in using GIS to detect the overlapping presence of relevant features to visually determine suitability.

Methodology

For this study, analysis of the Martian communications environment was performed using a geospatial model constructed in the QGIS program. First, the environmental parameters most affecting radio wave propagation were determined from the existing literature. They are:

1. Conditions of the ionosphere

2. Atmospheric conditions

3. Dust distribution and dust storm frequency

4. Major surface features (canyons, volcanoes, valleys, cliffs, etc.)

5. Surface clutter/roughness

Data Collection

Due to study limitations, only existing GIS map layers were used for this preliminary model.

Ionospheric Conditions

The ionospheric and atmospheric maps that were included in this model were generated using the Mars Climate Database Web Interface. A global Total Electron Content (TEC) map, evaluated at 1000 km above surface level, represents the ionospheric parameter most discussed in radio communication literature. This map is defined in greyscale for easier readability in conjunction with other maps.

Atmospheric Conditions

The atmospheric conditions represented are water vapor volume mixture ratio depicted in bluescale and O_2 volume mixture ratio shown in greyscale. Both were created using the Mars Climate Database Web Interface for the tropospheric region of the Martian atmosphere. Both parameters were indicated by Kliore et al. (2002) as primary factors affecting atmospheric absorption.

Dust Conditions

Dust conditions are a crucial and unique environmental factor affecting Martian communication ability. To represent these conditions, the Mars Dust Index map created by Ruff and Christensen (2002) from Mars Global Surveyor Thermal Emission Spectrometer data was used in accordance with Kuzma et al. (2018). Since wind speed was cited by Shekh et al. (2021) as the most significant dust-related factor determining attenuation levels, a surface wind stress and pattern map generated at 2 m above the ground using the Mars Climate Database Web Interface was used to supplement the Dust Index map.

Geographic Factors

Each of these ionospheric and atmospheric properties, while having reference to geographical features, do not represent the geographical features themselves that affect Martian communications. Surface roughness, rock abundance, and the presence of features creating drastic changes in elevation (such as craters, mountains, and canyons) all significantly impact the degree of attenuation, scattering, and other adverse conditions experienced in surface-to-surface communications such as would take place between rovers and/or human crews.

To represent these factors, a MOLA Global DEM and Global Color Shaded Relief Map were retrieved from the USGS's *Astropedia* data cache. These maps provide visual information on major surface features and elevation changes on a global scale.

A Ruggedness Index raster terrain analysis was performed on the MOLA DEM to create an additional GIS layer depicting surface roughness. This was supplemented with an ITRM Rock Abundance map created by Christensen (1986), indicating the presence of surface rock clutter across the middle regions of the Martian globe. Together, these provide an adequate estimate of influential topographical features for this preliminary analysis.

GIS Construction

Each of these data layers were imported as raster layers into the QGIS program.[1] The Coordinate Reference System was set to the native Mars 2000 CRS and applied using on the fly projection. Georeferencing with four or more major topographical features per map was necessary to ensure proper map alignment within the program. An additional vector layer was created using the Context Camera CTX image mosaic of human exploration zones on Mars (Figure 12). This map contained some apparent errors, primarily unregistered zones where one name seemed to apply to both areas. As the purpose of this layer was more to indicate general areas of interest where communications might be used than to provide an accurate tour of Martian topography, this was considered acceptable. In such cases the areas were simply combined and labelled with the available name. The resulting vector laycr consists of colored and labeled polygons indicating areas of interest for future missions, in order to provide navigation aid and to guide the relevance of map analysis.

After alignment of these layers was complete, the transparency of each layer was adjusted to create a visual overlay depicting all of these factors. Coloring and contrast were also adjusted in several layers by means of the Style and Transparency tabs of the Properties Panel to better accommodate combined viewing. The Hue function in the Style tab was used to shade the TEC map light pink and the O_2 volume mixing ration map light green. The Style tab was also used to change the MOLA Global Color Shaded Relief map from a multiband map to a single-band pseudocolor map with blue shading for better color continuity between map overlays. The other parameters altered for each layer are summarized in Table 1.

Due to the large numbers of layers involved, however, total transparency through all layers could not be attained. In order to adjust to this so that the data of each of these layers would be equally visible in the final product, the layers were grouped according to surface, ionospheric/ atmospheric, or dust condition data, so that each layer could be toggled through and the map set to show each group of factors and their global patterns determining communication conditions. The three final maps resulting from this were saved and exported as PNG image files, using the QGIS print composer. Figures 1 through 3 depict the final exported products.

Results and Discussion

These maps, depicted in Figures 1-3 below, reveal large scale patterns present in environmental factors affecting communications ability on Mars. Using them, it is possible to determine the most significant negative propagation phenomenon that radio communications can expect to experience in different regions. They can also aid in activities such as defining communication zones.

---

1   Due to technical limitations, this study used QGIS version 2.18.28 'Las Palmas.'

Table 1

Raster Layer Parameters Altered from Default

Layer	Transparency	Contrast	Saturation	Brightness
MOLA DEM	--	--	--	--
MOLA Color Shaded Relief	25	100	--	--
TEC Map	--	100	--	-166
O_2 Map	50	100	--	--
H_2O Map	50	100	--	-161
Wind Stress and Pattern	70	100	--	--
TES Dust Abundance	50	100	--	--
Ruggedness Index	64	--	--	--
IRTM Rock Abundance	80	100	100	87

Note. Void boxes indicate that the default settings were used.

Figure 1
Mars Global Ionospheric/Atmospheric Conditions for Communications Analysis

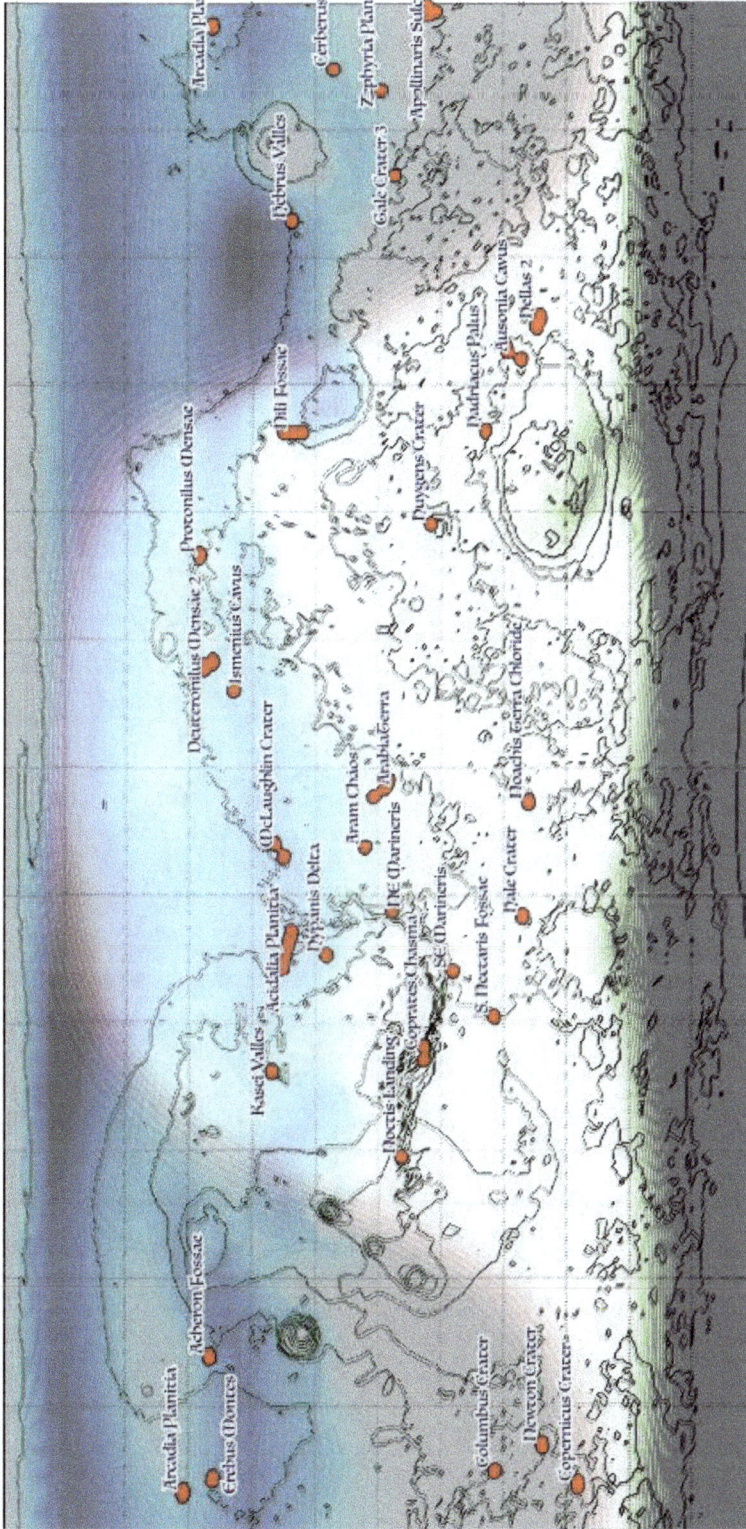

Note. Water Vapor Mixing Ratio represented by blue shading, darker blue indicating higher concentration. TEC represented by Pink lined shaded areas, and O2 Mixture Ratio represented by green shadowed areas.

Figure 2

Mars Global Dust Conditions for Communications Analysis

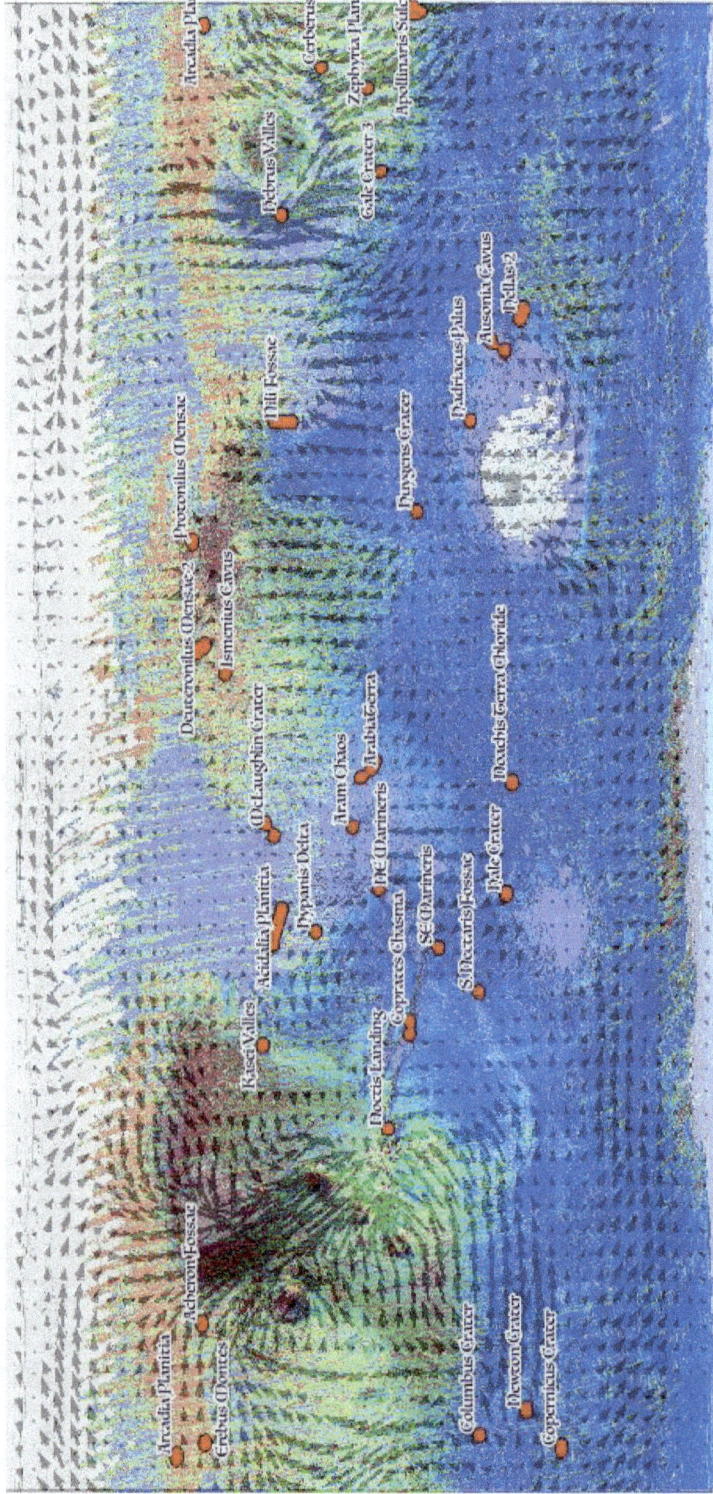

Note. Colored areas represent varying dust abundances, with warmer colors indicating greater abundances. Shaded areas indicate areas of greater wind stress, with arrows indicate wind pattern.

Figure 3

Mars Global Surface Conditions for Communications Analysis

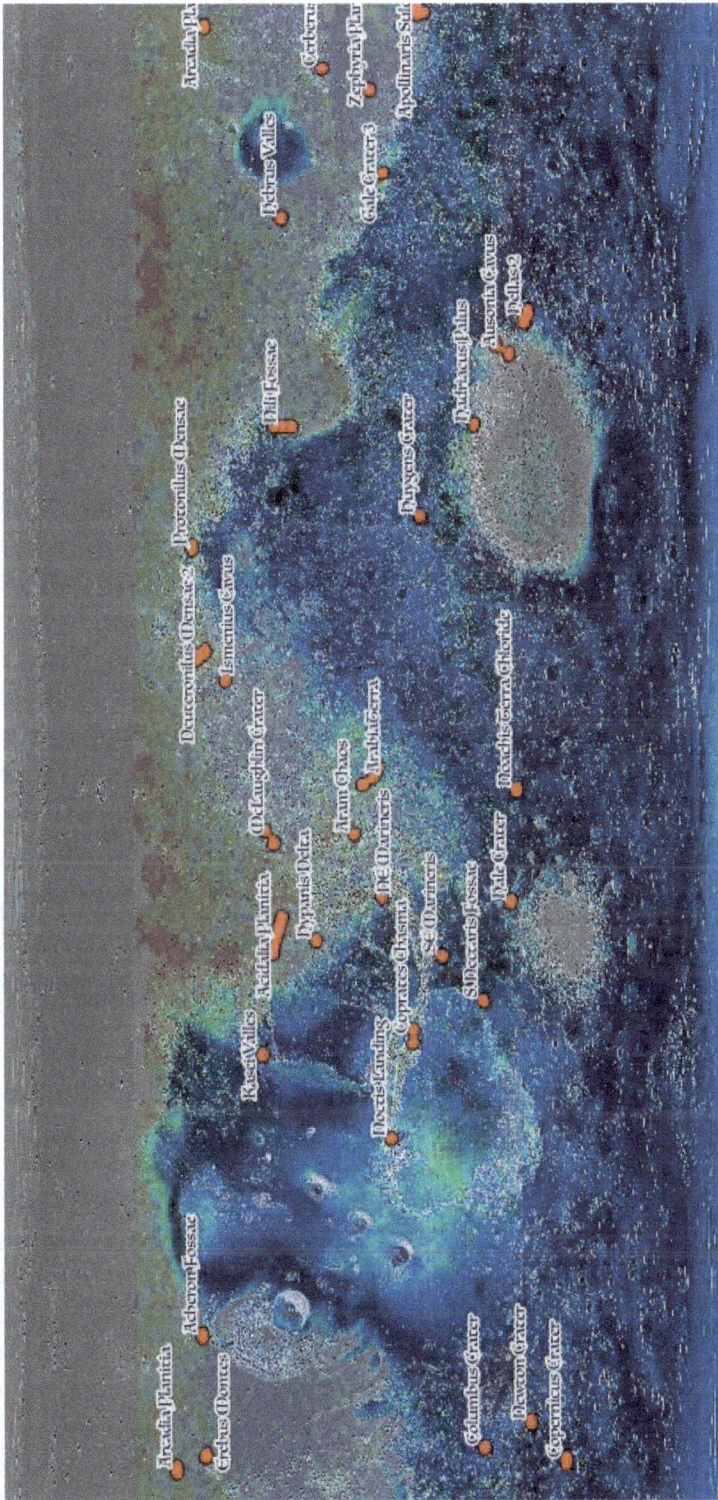

Note. Blue shading gradient indicates elevation, while color gradient represents rock abundance, with warmer colors indicating greater rock abundance.

Map Evaluation

Distribution of Negative Factors

Evaluation of these maps defines the dominating negative factors in particular regions as follows:

- The TEC of the Martian ionosphere can cause small but noteworthy Faraday rotation, range and time delays, and phase advance effects for vertical, one-way wave paths on signals below 450 MHz (Kliore et al., 2002). TEC effects are most significant around the northern pole and in the shaded region dipping down to the area of the Copernicus Crater to just east of the Hellas 2 exploration zone.

- The northern hemisphere and far southern hemisphere are susceptible to absorption in the microwave/infrared bands from H_2O and O_2 concentrations in the atmosphere in those areas. The areas in the middle of the map will generally be freer from these effects.

- Dust-related attenuation is dominant in the middle northern hemispheric regions above Olympus Mons and Elysium Mons, in the vicinity of exploration zones such as Arcadia Planitia, Kasei Valles, and Protonilus Mensae.

- Wind patterns are most prominent and erratic around the mountainous regions and the lower southern hemisphere.

- The northern regions above the mountains also may experience scattering effects due to the rockier terrain.

- Large scale scattering due to surface features such as craters creating rugged terrain is more dominant in the southern hemisphere and especially (and naturally) around the mountains.

Overall, the regions surrounding and northwest of Olympus Mons and northwest to northeast of Elysium Mons have a high concentration of potential negative factors, as well as the far southern region around the pole. Exploration zones such as Arcadia Planitia, Erebus Montes, Acheron Fossae, Protolinus Mensae, Deuteronilus Mensae 2, Nili Fossae, Hebrus Valles, and Cerberus fall within these more difficult areas. However, a large proportion of exploration zones inhabit the much freer middle zone of the maps. Here scattering due to rocky, rugged terrain is the likeliest contributor to attenuation, with few other majorly significant factors. The zone highlighted in purple in Figure 4 below shows this area featuring a low concentration of impeding factors.

Distribution of Positive Factors

Based on these maps it is also possible to identify regions of less impaired communications which are not presently marked for exploration. For instance, the northern parts of the Argyre and Hellas Basins suggest less potential for surface scattering (at least within their confines) and a general absence of other negative factors (see Figure 5).

Figure 4

Areas With a Lower Concentration of Potential Impediments

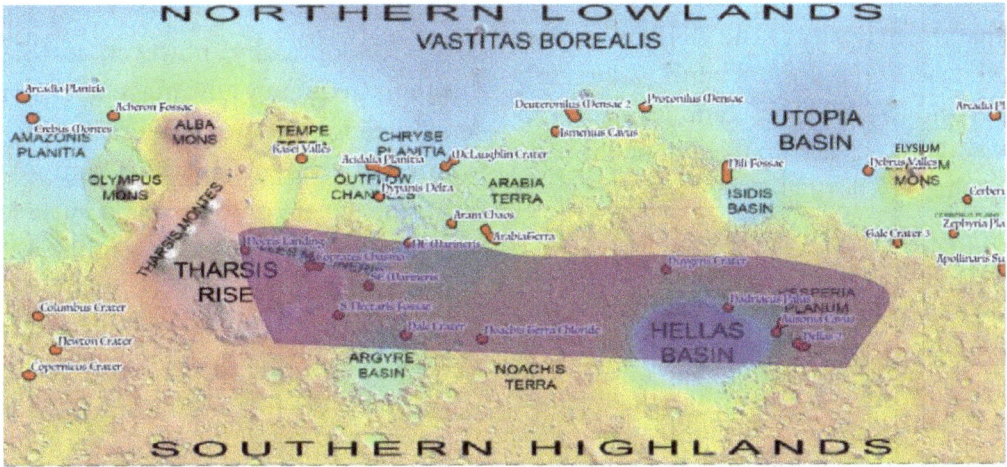

Note. Basemap from Mars MOLA experiment, NASA; modified by the author. Retrieved from https://www.planetary.org/space-images/map-mars-major-feature

---

Figure 5

Argyre and Hellas Basins Zones

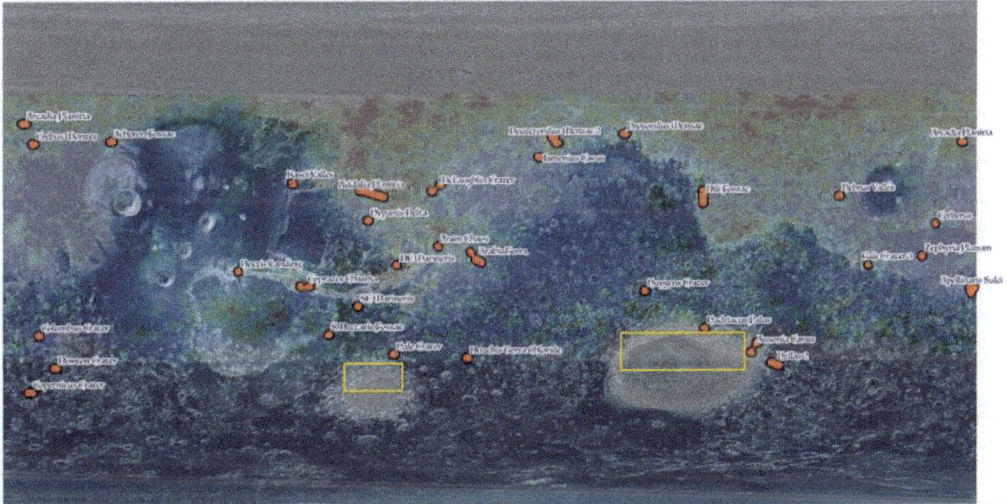

Note. Argyre and Hellas Basins Zones highlighted on the Figure 3 map generated for this study.

Relative Weight of Environmental Effects

When evaluating the maps for distribution of positive and negative effects, it is important to remember that the degree of these effects differ according to the different communication bands used. Therefore, not all effects will be equally weighted. As previously noted, the Martian ionosphere is incredibly thin, and signals greater than 450 MHz pass through unimpeded. This means that the majority of the UHF band (300 MHz – 3 GHz), the S band (2-4 GHz), X band (8 – 12 GHz), and Ka band (27 – 40 GHz) all are unaffected by the ionosphere (Kliore et al., 2002). However, lower frequencies (below 450 MHz) may experience increasing phase advancing, time delays, Faraday rotation, and range delay (Kliore et al., 2002). As ionospheric conditions fluctuate and are perturbed by varying solar conditions, absorption and bending also become more prominent in the lower frequency ranges (Kliore et al., 2002). Below 4.5 MHz, the ionosphere is impassible (Kliore et al., 2002). The atmosphere has a similar effect on different bands. Because it is so thin, most studies consider its direct effects negligible (Chukkala et al., 2004; Bonafini & Sacchi, 2020a; Kliore et al., 2002). Clouds, fog, or other aerosols are the primary attenuation concerns, and even these are estimated to contribute less than 0.3 dB in the Ka-band, and less for lower frequencies (Hansen et al., 2001; Kliore et al., 2002). Gaseous attenuation from H_2O and O_2 in the atmosphere may also occur (hence their inclusion in the maps); but because of their low concentration, the attenuation is generally on the order of 1 dB in the microwave band and none in the range of 60 GHz – 300 GHz (Kliore et al., 2002). Kliore et al. (2002) estimate this is increased by a factor of 1.5 at worst case conditions, which even so is not problematic. Overall, as long as communications stay within or above UHF range (pref-

erably 450 MHz or up), the ionosphere and atmosphere will not have a significant effect.

The difficulty, however, is that this high frequency advantage does not carry through the two remaining environmental elements affecting communications. Dust conditions and surface terrain are the two most significant impairments to Martian communications (Kliore et al., 2002). Dust-related attenuation due to dust storms can vary from 1 dB to a minimum of 3 dB for a very bad storm when the Ka band is employed (Kliore et al., 2002). Dust attenuation is determined by Hansen et al. (2001) to have a largely linear relationship with frequency, and lower frequencies experience lower attenuation. Likewise, multipath fading due to the ruggedness of the Martian surface is expected on the order of 2 – 7 dB within the UHF band (at 870 MHz). Higher frequencies, according to the model of Kliore et al. (2002), would experience even higher losses. This leads Kliore et al. (2002) to recommend low frequency bands for surface communications. In such a scenario, the ionospheric and atmospheric layers would have to be weighted with more consideration than they would be for high frequency bands. When considering high frequency communication, the ionospheric and atmospheric layers could be largely ignored, while the dust and surface layers would have a more critical status in the evaluation. Table 2 summarizes these relationships.

Table 2

Attenuation for Mars Communication Radio Frequency Bands

	VHF (100–500 MHz)	S-Band (2–4 GHz)	X-Band (10–12 GHz)	Ka-Band (30–38 GHz)
Ionosphere (absorption & scintillation)	0.5 dB	0.15 dB	0.1 dB	0.05 dB
Troposphere (scattering)	0	0	0	negligible
Gaseous	0	0 dB	0 dB	0 dB
Cloud	0	0	0.05 dB	0.1 dB
Rain	0	0	0	0
Fog	0	0	0	0.1 dB
Aerosol (haze)	0	0	0	0.1 dB
Dust*	0.1 dB	0.3 dB	1.0 dB	3.0 dB
Total Vertical Losses	0.5 dB	0.45 dB	1.15 dB	3.35 dB

* Worst case

Note. Reprinted with permission from *Radio Wave Propagation Handbook for Communication on and Around Mars* by Kliore et al., 2002, JPL Publication 02-5, 91.

Contextualizing Proposed Networks

Identifying areas of good or bad communication potential in this manner can help define areas for establishing initial networks to aid the development of Martian communications for future missions. It can also help coordinate mobile relays by identifying clear communication paths. Existing localized models such as those proposed by Chukkala et al. (2004), Borah et al. (2005), and Sacchi & Bonafini (2019) might benefit from contextualization within the larger patterns of the communication environment. The proposed networks largely operate within high frequency ranges. Chukkala et al. (2004) suggest UHF, S, and X band operations. Similarly, Sacchi & Bonafini (2019) propose an LTE-M network adapted from terrestrial LTE networks operating in the UHF band, with X, Ka, or higher bands used for long haul and

backhaul communications. The standard frequency for evaluation across these studies is UHF band communications at 2.4 GHz (Borah et al., 2005; Daga et al., 2006; Chukkala et al., 2004; Chukkala et al., 2005; Chukkala & De-Leon, 2005; Sacchi & Bonafini, 2019; Bonafini & Sacchi, 2020a; Bonafini & Sacchi, 2020b; Bonafini & Sacchi, 2021). The results of these studies generally reflect the propagation difficulties high frequency bands experience on the Martian surface due to abundant topographic changes and surface clutter. Chukkala et al. (2004), for instance, report a coverage of less than 50% in the Gusev Crater region, although they find this sufficient for basic communication demands. Bonafini & Sacchi (2021) find an average outage probability of 32% at 2.5 GHz in the rockier of their two Gale Crater sites, increasing with frequency. They report concern over the projected signal detection abilities in rockier sub-areas, which "become really scaring ... [and] almost problematic" (Bonafini & Sacchi, 2021, p. 8). Bonafini and Sacchi (2020b) report pathloss values of 60 dB or greater beyond 10 m distance between transmitter and receiver, and average shadowing across the two Gale Crater sites as being between 10 and 15 dB.

These site-specific scenarios can be expanded using the maps generated in this study (Figures 1-3) to evaluate potential global or regional communications networks following these signal parameters. From them, it is possible to discern on a general level which areas will see the most difficulty in communications. At 2.4 GHz, the ionospheric and atmospheric layers will be negligible, as the signal will be at a high enough frequency to pass through them largely unimpeded. The dust and surface condition layers, however, must be significantly weighted, as these conditions will more strongly impact large-scale signal performance. The regions around and to the north-northwest of Olympus Mons and the Tharsis Montes, as well as to the north and northwest of Elysium Mons, in the Utopia Basin and Arabia Terra, will be subject to significant dust attenuation (see Figure 2). Effects such as pathloss and multipath fading will be stronger in the rockier terrain within the Utopia Basin, Chryse Planitia, and more southern regions of the Vastitas Borealis, where the red and yellow areas are more prominent in Figure 3. The crater-pocked southern regions and the rugged terrain marking the descent of the highlands into the Northern Lowlands (the gray speckled, textured ar-

eas on the map in Figure 3) also pose significant potential propagation difficulties. The Northern Lowlands on the other hand, in those months when not covered by the polar ice cap, seem to pose less of a difficulty in terms of negative surface-related effects (see Figure 3). If low frequency communications were employed instead of the proposed network schema, however, the issue of surface ruggedness might not hold as much weight as it does for high frequency communications. In this scenario the ionospheric and atmospheric regions would need to be more carefully considered. The northern regions and the shaded "dip" west of Olympus Mons and east of Elysium Mons, as well the southern regions represented in green shadow in Figure 1 would see a greater degree of negative signal effects such as Faraday rotation, time and phase delay, absorption, and attenuation due to their higher TEC, H_2O, and O_2 content. These would carry more weight in evaluating areas for low frequency communication potential.

Limitations and Recommendations for Further Study

Being able to visualize and identify the patterns of communications-relevant environmental factors is useful in preliminary evaluations of the conditions in which proposed global or regional Martian communications networks must operate. While localized simulations can provide more specific data about a given area of interest, they do not provide ata-glance overviews of collected global scale parameters. This model enables prediction of large-scale communications patterns. Moreover, when in its native GIS environment, this map database provides the ability to toggle layers so that only those most significant to a given evaluation may be viewed. By overlaying this map data, the relevant factors can be viewed in their topographic relation to each other and areas of interest. This is an important foundation from which global network research may build, as it enables visualization of dominating factors in relation to each other.

However, this foundation is a not a comprehensive one. Both the detail and specificity of a localized model are lost within its generalized nature. It is not designed to provide specific evaluations with any high degree of accuracy. Instead, it provides a general backdrop of the global communications environment affecting factors. However, it by no means covers all of them. Because this study was limited to pre-existing and available GIS layers, several important factors are absent, most notably dust storm frequency (as opposed to the TES dust index, which portrays dust abundance).

Dust optical depth is likewise excluded. However, while this study does not currently include all of these, it should be noted that due to its nature it can be relatively easily expanded to show more. Given more time and research, for instance, an optical depth layer might be created. Suggestions for future work include incorporating the information available in the Mars Dust Activity Database (Battalio & Wang, 2019). This data exists in a CSV file format, which can be translated and uploaded as a layer in order to more effectively depict Martian dust storm frequency. This parameter is a crucial aspect of communications ability on Mars, but due to lack of existing data layers was omitted from this study. A vector layer portraying Martian satellite ground paths may also be a beneficial expansion for future study. There is also a possibility that variations in the environment might be modeled with regards to time as well as geography, which would to some extent reduce generalization.

Beyond additional data layers, the next step in terms of method would be to perform a suitability analysis of the resulting maps. This is difficult in the current format due to raster compatibility issues, but it would yield results from the map in a more practical, concrete form.

Each of these advances would increase the potential application of this preliminary global communications model, while further coverage pattern simulations could provide more specific data and might even be linked into their given areas in this model, for purposes of contextualization.

Conclusion

In this way, the maps developed in this study serve as a crucial steppingstone in the pursuit and development of a sufficient model of the Martian communication environment. While the majority of existing studies focus on localized modeling using conventional radio propagation simulation software, these approaches tend to limit the study in scope. Due to the expected range and scope of future Mars exploration and the essential nature of communications to these space missions, it is important to prepare a method of understanding the Martian communications environment on a global scale. This will allow for the more successful and secure planning and implementation of global communications networks on the Martian surface and immediately within its vicinity. GIS, as a widely available application specializing in the synthesis of bulk geographically-related information, represents an ideal and convenient starting place in the effort to compile a viable global model of the communications environment on the Red Planet. The maps of the ionosphere/atmosphere, dust-related, and surface factors related to communication effectiveness that were created in this study show pertinent, if generalized, large-scale environmental patterns. These can allow the viewer to determine both the dominating potential impairments in a region and the areas in which communications would likely be most impeded by environmental factors. Such a model readily allows for further additions. By building on this model, future researchers can develop a clear, integrated, and comprehensive understanding of the potentials of the Martian communications environment as a whole. Research built on this model can help manage expectations regarding what kind of frequencies, equipment, and networks would best be capable of handling the environmental parameters involved, so that any future global networks for Mars may be designed to be as reliable as possible. It is towards this critical goal that this study hopes to contribute, so that humanity may remain connected no matter the lightyears between.

References

Battalio, J. M., & Wang, H. (2019). The Mars Dust Activity Database (MDAD). Harvard Dataverse. doi:10.7910/DVN/F8R2JX.

Bonafini, S. & Sacchi, C. (2020a). Building cellular connectivity on Mars: A feasibility study. *2020 IEEE Aerospace Conference.* doi: 10.1109/AERO47225.2020.9172518.

Bonafini, S. & Sacchi, C. (2020b). Evaluation of large scale propagation phenomena on the Martian surface: A 3D ray tracing approach. *2020 10th Advanced Satellite Multimedia Systems Conference and the 16th Signal Processing for Space Communications Workshop (ASMS/SPSC).* doi: 10.1109/ASMS/SPSC48805.2020.9268846.

Bonafini, S., & Sacchi, C. (2021). 3D ray-tracing analysis of radio propagation on

Mars surface. *2021 IEEE Aerospace Conference.* doi: 10.1109/AERO50100.2021. 9438180.

Borah, D.K., Daga, A., Lovelace, G.R., & De Leon, P. (2005). Performance evaluation of the IEEE 802.11a and b WLAN physical layer on the Martian surface. *IEEE Aerospace Conference Proceedings.* 1429 - 1437. doi:10.1109/AERO.2005.1559433.

Christensen, P.R. (1986). The spatial distribution of rocks on Mars. *Icarus 68* (2): 217-238. doi: 10.1016/0019-1035(86)90020-5.

Chukkala, V., De Leon, P., Horan, S., & Velusamy, V. (2004). Modeling the radio frequency environment of Mars for future wireless, networked rovers and sensor webs. *2004 IEEE Aerospace Conference Proceedings 2* (IEEE Cat. No.04TH8720). doi: 10.1109/AERO.2004.1367731.

Chukkala, V., DeLeon, P., Horan, S., & Velusamy, V. (2005). Radio frequency channel modeling for proximity networks on the Martian surface. *Computer Networks 47*: 751-763. doi:10.1016/j.comnet.2004.08.011.

Daga, A., Borah, D.K., Lovelace, G.R., De Leon, P. (2006). Physical layer effects on MAC layer performance of IEEE 802.11 a and b WLAN on the Martian surface. *Aerospace Conference, 2006 IEEE.* doi: 10.1109/AERO.2006.1655780.

Hansen, D.M., Sue, M.K., Ho, C.M., Connally, M., Peng, T.K., Cesarone, R.J., & Home, W. (2001). Frequency bands for Mars in-situ communications. *IEEE Aerospace Conference Proceedings 3*, 3/1195 - 3/1208. doi: 10.1109/AERO.2001.931350.

Kliore, A., Golshan, N., & Ho, C. (2002). *Radio Wave Propagation Handbook for Communication on and Around Mars.* (JPL Publication 02-5). NASA.

Kuziakina, M., Gura, D., & Zverok, D. (2019). GIS analysis of promising landing sites for manned flight to Mars. *E3S Web Conference 138* (02004). https://doi.org/10.1051/e3sconf/201913802004

Kuzma, M., Gładysz, Ł., Gralewicz, M., & Krawczyk, P. (2018). Applications of GIS in analysis of Mars. [Conference proceedings]. *Proceedings of GIS Ostrava,* 267-276. doi: 10.1007/978-3-319-61297-3_19.

Ruff, S.W., & Christensen, P.R. (2002). Bright and dark regions on Mars: Particle size and mineralogical characteristics based on Thermal Emission Spectrometer data. *Journal of Geophysical Research, 107.* doi: 10.1029/2001JE001580.

Shekh N.A., Dviwedi V., & Pabari J.P. (2021). Effect of sandstorm on radio propagation model of Mars. In: Raj, J.S. (eds) *International Conference on Mobile Computing and Sustainable Informatics.* ICMCSI 2020. EAI/Springer Innovations in Communication and Computing. Springer, Cham. doi:10.1007/978-3-030-49795-8_43

USGS Astrogeology Science Center. (n.d.) *Astropedia.* https://astrogeology.usgs.gov/search?pmi-target=mars

Seizing the Stars: Resources, Expansion, and Counterspace Contingencies Across the Space Domain

Joshua E. Duke

Author Note: This first appeared as "Conflict and Controversy in the Space Domain: Legalities, Lethalities, and Celestial Security," in *Wild Blue Yonder*, the DAF Air University blog, on 29 September 2020, and has been updated by the author prior to republication. (https://www.airuniversity.af.edu/Wild-Blue-Yonder/Article-Dis play/Article/2362296/conflict-and-controversy-in-the-space-d omain-legalities-lethalities-and-celesti/)

Abstract

Space is becoming the next frontier for human conflict and competition. The United States, the People's Republic of China (PRC), and the Russian Federation (RF) have all invested deeply in a modern space race to gain or maintain strategic superiority, with plans for lunar bases, celestial resource exploitation, and the colonization of Mars. With technological advancements and a weak regulatory framework governing space operations, the development of space-based and counterspace military assets, advanced space weaponry, space transportation and space resource exploitation operations are an inherent part of mankind's future. This article assumes the inevitability of space exploration—including celestial body resource exploitation, weapon research and developments, and the human colonization of Mars—to show the importance of American leadership of human expansion into space. Power in space will be drawn from technological developments, including new types of weaponry and energy production. The author explores the technologies available in today's space race environment, including potential future energy resources available in space, weapon systems designed for space and counterspace warfare, the legal implications of each, and some potential consequences of different nations gaining the upper hand in the heavens.

Keywords: space, counterspace, ASAT, lunar exploration, space resources, fusion power, China, Russia, united states, international

doi: 10.18278/sesa.4.1.3

space law, helium-3, Mars, anti-satellite, space warfare, weaponization, technology, quantum communications, space exploration, moon, commercialization

Aprovechando las estrellas: recursos, expansión y contingencias contraespaciales a través del dominio espacial

Resumen

El espacio se está convirtiendo en la próxima frontera para el conflicto y la competencia humanos. Estados Unidos, la República Popular China (RPC) y la Federación Rusa (RF) han invertido profundamente en una carrera espacial moderna para ganar o mantener la superioridad estratégica, con planes para bases lunares, explotación de recursos celestiales y la colonización de Marte. Con los avances tecnológicos y un marco regulatorio débil que rige las operaciones espaciales, el desarrollo de activos militares basados en el espacio y contraespaciales, el armamento espacial avanzado, el transporte espacial y las operaciones de explotación de recursos espaciales son una parte inherente del futuro de la humanidad. Este artículo asume la inevitabilidad de la exploración espacial, incluida la explotación de recursos del cuerpo celeste, la investigación y el desarrollo de armas y la colonización humana de Marte, para mostrar la importancia del liderazgo estadounidense en la expansión humana en el espacio. El poder en el espacio se extraerá de los desarrollos tecnológicos, incluidos nuevos tipos de armamento y producción de energía. El autor explora las tecnologías disponibles en el entorno de la carrera espacial actual, incluidos los posibles recursos energéticos futuros disponibles en el espacio, los sistemas de armas diseñados para la guerra espacial y contraespacial, las implicaciones legales de cada uno y algunas consecuencias potenciales de las diferentes naciones que obtienen la ventaja en los cielos.

Palabras clave: espacio, contraespacio, ASAT, exploración lunar, recursos espaciales, energía de fusión, China, Rusia, Estados Unidos, ley espacial internacional, helio-3, Marte, antisatélite, guerra espacial, armamento, tecnología, comunicaciones cuánticas, exploración espacial, luna , comercialización

抢星：跨越太空领域的资源、扩张与反空间突发事件

摘要

太空正成为人类冲突和竞争的下一个边界。美国、中华人民共和国（PRC）与俄罗斯联邦（RF）都在现代太空竞赛中投入巨资，以获得或保持战略优势，计划建设月球基地、开发天体资源以及殖民火星。鉴于技术进步和用于治理太空操作的监管框架薄弱，基于太空的反太空军事资产的开发、先进的太空武器、太空运输、以及太空资源开发操作是人类未来的内在组成部分。本文假设了太空探索的必然性——包括天体资源开发、武器研发以及人类殖民火星——以展示美国领导人类向太空扩张的重要性。太空实力将来自技术发展，包括新型武器和能源生产。作者探究了当今太空竞赛环境中的可用技术，包括太空中潜在的未来能源资源、为太空和反太空战设计的武器系统、每种技术的法律含义、以及在太空领域中占上风的不同国家的一些潜在结果。

关键词：太空，反太空，ASAT，月球探测，太空资源，核聚变能，中国，俄罗斯，美国，国际空间法，氦-3，火星，反卫星，空间战，武器化，技术，量子通信，太空探索，月球，商业化

"Adversary action in space is inevitable, and the adversary will generate effects that deny, degrade, and disrupt the space operating environment."

– United States Marine Corps Tentative Manual for Expeditionary Advanced Base Operations 2021: 75.

Introduction

Space is becoming the next frontier for human conflict and competition. The United States, the People's Republic of China (PRC), and the Russian Federation (RF) are the three most powerful nations on Earth, all of which have invested deeply in a modern space race to gain or maintain strategic superiority. Each of these nations has plans for lunar bases, celestial resource exploitation, and the colonization of Mars. The RF and PRC have even announced a partnership to develop a joint Moon base in response to the U.S.-led Artemis project, which includes the establishment of Artemis

Base Camp at the lunar South Pole (La Rocca, 2022, 34; NASA, 2020). All three nations have also developed, or are in the process of developing, a variety of counterspace weaponry, space-based weapon systems, and spacecraft capable of maneuvering in zero gravity, the combination of which can and will be used to control space and potentially the future of mankind. Existing international laws and treaties regulating space initiatives, notably the 1967 Treaty on Principles Governing the Activities of States in the Exploration and Use of Outer Space, Including the Moon and Other Celestial Bodies (Outer Space Treaty), lack sufficient legally binding language when applied to today's space-based technologies and concepts for developments (United Nations, 1967). There are few international recourses available under the existing regulatory framework to prevent a nation from developing space-based or counterspace military assets, to include weaponry, or to prevent the exploitation of resources in space.

This article assumes the inevitability of space exploration—including celestial body resource exploitation, weapon research and developments, and the human colonization of Mars—to show the importance of American leadership of human expansion into space. As national and international organizations are reducing barriers to entry, while increasing access to space-based activities, a hierarchy will inevitably emerge. Power in space will be drawn from technological developments, including new types of weaponry and energy production. Resource exploitation in space and on other planets will drive industry and economic development on Earth in exponential increments as humans expand into the space domain, driving the need for both a guiding force and international cooperation to avoid conflict. As information age generations look to the stars to answer their needs and ambitions, a parallel generation will emerge—a space age generation—with an eye to protect or to control, depending on who maintains the greatest portion of power in space. American leadership is critical as mankind explores the stars to ensure that both a freedom-centric ideology and free-market capitalism become the guiding tenets of space exploration.

The author explores the technologies available in today's space race environment, including potential future energy resources available in space, weapon systems designed for space and counterspace warfare, the legal implications of each, and some potential consequences of different nations gaining the upper hand in the heavens. Part 1 outlines recent space-relevant technological developments. Part 2 examines lunar exploitation and resources, particularly Helium-3, and the potential for future fusion energy developments. Part 3 explores the potential benefits of exploring, exploiting, and colonizing Mars. Part 4 underscores the severity of the potential and actuality of space weaponization, including an overview of existing and theoretical weaponry and legal implications. Finally, Part 5 concludes with an analysis of the potential implications of recent developments and control over space and celestial bodies with regard to global economic stability and space superiority, emphasizing the absolute need of American leadership as humans expand into the space domain.

1. Space Technologies and Advancements

Advancements in space technology are quickly leading to an inevitable conflict over control in space, which includes control over the Moon through lunar bases and potentially control over the colonization of Mars. The PRC has added several capabilities into its military space program, including "antisatellite [ASAT] interceptors, miniature space mines, and ground-based lasers" that can conduct attacks on other satellites (Hughes, 2011, 24). These capabilities fall under the guise of the Outer Space Treaty's permission to destroy militarized satellites (Pool, 2013). ASAT technologies can easily be used offensively to create a decision advantage in combat. Some analysts believe that the deliberate collision of PRC satellites with older satellites shows that the PRC has experimented with "parasitic satellites" designed to lie dormant in the vicinity of a target until activated, potentially for hacking or debilitating purposes (Hughes, 2011, 25-26). Robotic technologies on satellites have also been demonstrated, including robotic arms, which will likely lead to on-orbit ASATs "designed to hijack, jam, re-purpose, exploit, destroy or covertly monitor" adversary satellites (NATO, 2020, 81).

The PRC has two space planes in development, the Shenlong and Tengyun, and in 2020, they successfully launched a space plane prototype, which orbited Earth for two days before returning to the surface (Defense Intelligence Agency, 2022, 34). The PRC continues to be locked in an intense space race with Russia and the United States, with a short-term goal of controlling the Moon with a lunar base and a longer-term goal of populating Mars under the rule of the PRC (Hughes, 2011). The development of maneuverable space planes and lunar bases is not unique to the PRC. The National Aeronautical and Space Administration (NASA) developed the X-37 and X-37B space planes, and the Russian Federation is developing a maneuverable space plane using nuclear technology for power (Hughes, 2011). All of these nations are expanding their space activities drastically and have planned missions to the moon and Mars over the next 30 years (Defense Intelligence Agency, 2022, 40; NASA, 2020). The nation that achieves these goals first will be positioned to set the standards for life and activities on celestial bodies, be it democracy or dictatorship.

Despite the array of international treaties and agreements promoting peaceful global development of space resources in the name of science and humanity, it is unlikely that space will remain weapon free and likely that it will become the next frontier of global combat. Space weapons in use and under development may use robotics, nanotechnology, cyber weapons and directed energy such as microwaves and lasers (Jensen, 2014). With the establishment of a lunar base, a nation with advanced laser technology, advanced cyber weaponry, maneuverable space planes, satellite targeting capabilities, nano-science stealth technology, artificial intelligence, quantum communications, and self-guiding nanotechnology bullets would undoubtedly have the capacity to rule the Earth as it sees fit. All of these technologies already exist or are in development phases, and they are the future of intelligence and warfare (Jensen, 2014; NATO, 2020). Additionally, the U.S. government and NASA have been encouraging the commercialization of space cargo transportation to meet future American needs for access to the International Space Station (ISS) and to improve the research and development of spaceborne technologies and other developments, most recently through the announcement of the Artemis program (Hughes, 2011; NASA, 2020).

Private sector involvement has opened the market for alternative rocket propulsion technologies that can achieve government and commercial goals for space at lower costs and faster than possible under the existing bureaucracy of NASA. Enhanced private sector involvement in space travel utilizes the free-market system to foster radical developments and investment for both government and private sector programs, incentivizing broader participation, which benefits both. The PRC's communist version of capitalism is also expanding commercialization of space activities, but with authoritarian leadership all private sector in-vestments, including technologies and other space-related innovations, directly benefit the PRC government. Commercializing aspects of standard space operations, such as recent and planned operations involving SpaceX and Blue Origin, will reduce barriers to space over time, including lowering costs, normalization of space tourism, introduction of a space transportation industry, and extraterrestrial resource exploitation activities, particularly asteroid mining and celestial body mining operations. Commercialization of space operations will free up resources for NASA and the newly minted U.S. Space Force to pursue broader goals, such as manned deep space travel, lunar-based activities, and manned missions to Mars.

2. Lunar Power

Rare earth metals and other minerals are quickly becoming scarce in the United States to the point where the international space race to claim the Moon and Mars has become a top priority, not just for control, but for resources available for exploitation. Uranium has even entered the economic radar as a good idea for boosting the American economy instead of remaining too dangerous to mine due to the associated health risks and environmental hazards. Uranium is in abundance on the Moon (Crawford, 2015). Estimates suggest there may also be up to five million tons of Helium-3 (^{3}He) contained within the lunar regolith (Dobransky, 2013). This has the potential to meet all of mankind's power needs for thousands of years when used for fusion power (Dobransky, 2013). On top of the resources potentially available, the Moon provides a unique launching position for future missions to Mars with a faster, more direct, and more efficient path to the Red Planet (Dobransky, 2013). Control over the Moon is an inherent factor in the future of energy production, strategic power, and the human race.

Uranium has long been a part of the nuclear fission enterprise on Earth but comes with high costs, including radioactive waste and extreme health and environmental hazards due to the radiation produced in the fission pro-cess. Terrestrial reserves of other energy-producing resources, like oil and natural gas, have also been projected to be exhausted within 50–100 years under current and projected mining and usage rates (Dobransky, 2013). Alternatively, the element tritium (T), which has a half-life of 12.32 years, naturally decays into ^{3}He (Kolasinski, Shugard, Tewell, and Cowgill, 2010, 5), which can be used to create a new kind of power—fusion power. Fusion power can be generated by combining deuterium (D) with either more D, T, or ^{3}He, using the following calculations shown in order of their ignition temperatures:

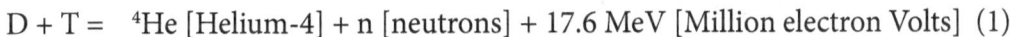

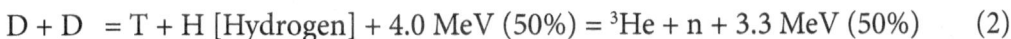

$$D + T = \quad ^4He \text{ [Helium-4]} + n \text{ [neutrons]} + 17.6 \text{ MeV [Million electron Volts]} \quad (1)$$

$$D + D \quad = T + H \text{ [Hydrogen]} + 4.0 \text{ MeV } (50\%) = ^3He + n + 3.3 \text{ MeV } (50\%) \quad (2)$$

$$D + {}^3He \quad = {}^4He + H + 18.4 \text{ MeV (Hughes, 2011)} \quad (3)$$

Fusion power can also be created by combining ^{3}He with more ^{3}He, creating Helium-4 (^{4}He) (Dobransky, 2013). The combination of ^{3}He and ^{3}He is the most energy efficient, producing the greatest net energy, but also requires the highest ignition temperature to achieve fusion (Dobransky, 2013; Crawford 2015, 157). Fusion power generation using ^{3}He produces the cleanest and most abundant energy, but is also the most difficult to achieve.

Unfortunately, ^{3}He exists only in minute amounts on Earth (Dobransky, 2013). The nation that establishes a mining and transportation industry capable of bringing lunar ^{3}He to Earth, and develops a fusion plant network that transforms ^{3}He into power, could control a substantial portion of the planet's energy industry for decades. Some scientific estimates discount both the estimates of the potential amount of extractable ^{3}He in the lunar regolith and the potential to achieve industrial fusion reactors on Earth capable of processing it. Exemplifying this scientific stance are the calculations of Ian Crawford, who believes both prospects are greatly exaggerated and that there are only approximately 220,507 tons of ^{3}He available in logical extraction areas, such as the titanium-rich lunar basalt flats (Crawford, 2015, 144-145). Despite his dissent, Crawford admits even lunar resources that seem impractical and economically inefficient to transport resources to Earth may provide substantial economic benefits for space-based uses, such as solar power systems and spacecraft fusion engines, which would not require transport back to Earth (Crawford, 2015, 145).

The concept of fusion power has gone through several stages of the Gartner Hype Cycle over many decades, beginning in the middle of the twentieth century when the concept was first being explored, followed by decades of disillusionment after the initial peak of enthusiasm (NATO, 2020, 11-12). Modern technological advancements and increased commercial interest, combined with investment in space exploration activities, have moved the idea of fusion power into the slope of enlightenment as proofs of concept have begun to multiply, and as extraterrestrial resources that can be used in creating fusion power have become targets of opportunity in space for both nations and private investors. Earth's finite resources make lunar and space resource exploitation an inevitability. The most pertinent factor governing future human resource exploitation in space is the question of which nation will achieve a successful and effective industrial supply chain first. The most probable three nations to achieve this are the U.S., the PRC, and the RF, and the three areas that need to be navigated to succeed are facility establishment, production/refinement, and transportation.

Establishing lunar facilities is the easiest of these goals, especially when lunar resources that can be used for building are taken into account, which decreases the amount of materials needed to be brought to the Moon and the time needed for construction. In 2008, a NASA experiment found that lunar regolith has potential construction properties. When scientists heated the regolith and used sulfur as a binding agent, they made "waterless concrete," which can be molded and is nearly as strong as concrete when it hardens (Hughes, 2011, 45-46). This process requires minimal effort and relies primarily on direct heat application and the ability to shape the regolith. Consequently, the entire process can be automated by robots with the appropriate tools on the lunar surface, such as the ones NASA began developing specifically for this purpose in 2009 (Hughes, 2011). The simplicity of the operational requirements means that these three nations already have the technical capability to begin construction using lunar soil after arriving on the Moon. They will also all be capable of bringing any other materials that would be necessary to construct facilities or bases on the lunar surface.

Unlike the U.S., and contrary to existing international law, the PRC's stance on the Moon is that it is territory, despite the prohibition on "national appropriation" of celestial bodies outlined in Article II of the Outer Space Treaty (United Nations 1967) (Hughes, 2011). The PRC has also proposed mining ^{3}He for future fusion power opportunities (Hughes, 2011). The RF, while not openly pursuing a territorial ambition for the Moon, is exploring and advancing prospects of economic development, including ^{3}He extraction

(Hughes, 2011). Firms in several countries, including the United States, Great Britain, Japan, and Russia, are also developing spacecraft for tourism, which will inevitably improve technologies useful for other purposes, including space cargo transportation (Defense Intelligence Agency, 2022, 35). Facility development and resource exploitation areas on the Moon are limited. This will exacerbate the race for prime locations and desirable resources, particularly at the poles, where water ice is believed to exist in large quantities (which can be used to sustain lunar human habitation), and in the titanium- and ^{3}He-rich basalt flats of Mare Tranquillitatis and Oceanus Procellarum (Crawford, 2015, 145). Once established, facility operations, such as the planned Artemis Base Camp at the lunar South Pole, can begin to extract and refine resources either for use on the lunar surface or for transportation to Earth (NASA, 2020).

Transportation of materials from the Moon to Earth is a substantial financial and logistical undertaking, and it will not be easy to show a profit after the considerable expenses associated with it. Nevertheless, extraction and transportation of ^{3}He and other resources to Earth, specifically for fusion power production, have been expressed as long-term goals of the PRC and the RF within the next decades. Interestingly, the U.S. has not stated this as a specific goal, but it has already shifted its space transportation industry sufficiently toward the private sector to achieve it, while initiatives of the Artemis program include resource exploitation activities (NASA, 2020, 28-29, 61).

U.S.-based private sector organizations will have the most viable opportunity to build the first industrial space transportation system, specifically because of advantages in the American free-market system (Hughes, 2011). By encouraging private sector participation in the space industry and commercializing space transportation, the U.S. has also made production of space technologies competitive with proposals in the National Space Policy (Obama, 2010, 3-5). A competitive industry makes substantial investments in research, development, and production of space transports; engine components for space travel; and tools for use in zero gravity. America cannot afford to fall behind in the race for lunar facility establishment and resource exploitation, to maintain economic and national security, and to secure the future of human expansion into space, as the Moon offers the most efficient launching position for missions to Earth's red neighbor, Mars.

3. Mars Domination

Mars is widely accepted by the scientific community to be the most plausible planet for the first human habitation on a celestial body and, consequently, the most likely location for the first space colony and eventually a second planet for humankind. Thus, Mars is a desirable goal for nations involved in space exploration for many reasons, which the United States plans on pursuing with humans landing on the Red Planet for the first time in the 2030s (NASA, 2020, 59). The territory on Mars will also most likely become marketable for economic value to civilians in the long term, in addition to resource exploitation activities. The Outer Space Treaty prevents ownership of territory on celestial bodies but makes no mention of ownership or sale for profit of structures built on, or items brought to, celestial bodies, just as there is no explicit language in the treaty preventing profit-based resource exploitation on celestial bodies by either governments, organizations, or private nationals (United Nations, 1967).

The inevitability of Mars becoming a second planet inhabited by humanity must be considered, along with all of the implications of living spaces and ownership of property that will eventually follow. Denying this inevitability and claiming it as outlawed by international law due to the prohibition on appropriating territory on a celestial body would essentially equate owning property on Earth as also outlawed by international law. After all, Earth is also a celestial body. Language in the treaty encourages expansion into space and essentially says that if persons, governments, or organizations build something on a celestial body, they own that building and can do what they want with it, including selling it (United Nations, 1967). They cannot, however, claim to own the planet's ground outside the building—yet. Resources on Mars, while still not mapped out as substantially as lunar resources have been, will likewise create new markets for economic prosperity and national wealth, including more ^{3}He deposits from solar winds like those found in lunar regolith, along with substantially

high concentrations of iron (Dobransky, 2013).

In addition to buildings constructed on celestial bodies, spacecraft and facilities constructed in space and on celestial bodies are also considered to be the territory of the owning nation, which means that the UN Charter applies to facilities and spacecraft in space and on celestial bodies. UN Charter Article 2(4), in particular, protects space explorers and potential future residents on Mars by prohibiting the "use of force against the territorial integrity" of another nation party to the treaty (United Nations, 1945), which all space-faring nations are. Article 51 further dictates that if attacked, "the inherent right of ... self-defence" shall not be impaired (United Nations, 1945). Article V of the Outer Space Treaty prescribes that, in space, all humans are bound to "render all possible assistance to" each other as "envoys of Mankind" (United Nations, 1967). Essentially, a peaceful international course is possible—even mandated—for human expansion into space. Unfortunately, the PRC and the RF regard space and celestial bodies as territorial goals, leading to the assumption that attempts will be made to control and defend such territories as necessary to achieve space superiority, control over space resources, and managerial power over the future colonization of Mars (Hughes, 2011).

Control over Mars, in addition to affecting resource exploitation, transportation, and scientific advancements, also has implications for the direction of humanity in space. Establishment of a human colony, or human colonies, on

Mars will eventually lead to territorial spaces, development of the land and air (potentially involving terraforming the planet for atmospheric enhancement), and security issues. While an established colony on the Red Planet is still likely decades away, trends within the PRC and RF governments suggest that any established colony on Mars under their jurisdiction would be authoritarian, weaponized, and secret. Given the nature of weather on Mars, fortified structures are easily justified, and the lack of a conventional weapons ban on celestial bodies makes weaponization of such a colony both legal and desirable, mainly because of the third inherently desired factor—secrecy. The inevitability of PRC and RF presence on Mars also suggests that any U.S. developments will likely include fortifications and weaponization. While the Outer Space Treaty mandates cooperation between nations on celestial bodies, the extreme distance between Earth and Mars means that a compliance verification system with effective monitoring and enforcement will be complicated, if not impossible, for the foreseeable future. For these reasons, a nation that effectively controls near-Earth space and establishes a security presence on the Moon will effectively be in a position to control Mars.

4. Space and Counterspace

Celestial bodies are not the only potential fields of conflict in space, and in the short term, space itself has become a much more immediately relevant focus for space-faring nations and the world. This is particularly the case in the vicinity of Earth, including orbital paths for communication technologies, weapon platforms, and sensors. Technological improvements and the proliferation of nation-state and private sector interest and capacity to enter space are causing the acceleration of an inevitability—usable orbital space around Earth is diminishing (Koplow, 2014). Satellites and other spaceborne assets orbiting Earth are quickly filling up all of the most useful places to perform their assigned functions within Earth's various orbits, and space debris is complicating matters even further. Increasing numbers of space objects are causing difficulty in establishing safe orbital paths for newly launched spacecraft while increasing the risk to launches destined for deep space (Chanock, 2013). Adding to these complications are international developments of ASAT weapons, many of which add to the approximately 100 million pieces of space debris traveling as fast as 17,500 mph already orbiting Earth (Garcia, 2021; Koplow, 2014, 796-797).

ASATs in use and under development, with attacks initiated using space-based, ground-based, and airborne delivery methods, include essentially three broad areas: kinetic energy (KE), such as missiles, rail guns, or other satellites impacting targets in space; directed energy (DE), which includes lasers and particle beams; and electronic/cyber weapons (Koplow, 2014, 795; Koplow, 2019, 305-306). Counterspace weapons include three categories: space to space, which includes satellites targeting other satellites; space to ground, such as satellite weapons targeting Earth; and ground to space, encompassing ground launched weapons targeting satellites (Harrison, 2021, 3). The Outer Space Treaty, while prohibiting nuclear weapons from being used in any way in space including being stationed in space, "has no specific provision prohibiting the use of conventional weapons, [including lasers], in outer space" (Jensen, 2014, 275), which inherently authorizes them. The Outer Space Treaty also contains no prohibition of such weapons being stationed on space-based platforms, including on celestial bodies, or of them being used to target objects on Earth, in space, or on celestial bodies (Jensen, 2014). In other words, these weapons are legal in every way, regardless of the potential damage they can cause to international stability and humanity.

There are multiple ongoing debates over the nature, definitions, and classifications of several kinds of ASATs currently in operation or in developmental phases. With NATO's 2021 declaration that Article 5 could be invoked in the event of "attacks to, from, or within space," counterspace and counter-counterspace activities have become a very high priority for understanding, developing, and fielding (Calcagno, 2022, 37). Space to space ASATs include several types of satellites designed to initiate operations in close proximity to adversary space assets for purposes of "inspection, manipulation, damage, or capture" (Koplow, 2019, 305). Orbital satellites with robotic arms can also potentially launch cyber-attacks against other satellites to destroy, disable, or control them through direct attachment, parasitically (Defense Intelligence Agency, 2022, 18). Space-based ASATs, such as the prototype Russian ASATs Cosmos 2504 and Cosmos 2536, can also directly impact other satellites to cause kinetic damage (Defense Intelligence Agency, 2022, 29). The RF and the PRC have also fielded several ground-based ASAT systems, including missiles and ground-based lasers that can be used to blind sensors, damage components, or incapacitate satellites (Defense Intelligence Agency, 2022, 17, 28).

Space is a warfighting domain according to Russia, and they have developed missiles designed to destroy assets in space, including space vehicles and satellites, with little regard for creation of space debris (Defense Intelligence Agency, 2022, 21). In November of 2021, Russia tested its Nudol ASAT weapon system, creating over 1,500 trackable pieces of debris, and "tens of thousands of pieces of lethal but nontrackable debris," endangering all spacecraft in Low Earth Orbit (LEO) (Defense Intelligence Agency, 2022, 28). Russia is also "reportedly developing an air-launched ASAT weapon called Burevestnik" that can target spacecraft in LEO (Defense Intelligence Agency, 2022, 29). The PRC and the RF continue advancements in ASAT research and development, fielding new counterspace weapons on a regular basis to hold U.S. and allied space capabilities at risk, including kinetic, directed energy, and cyber-attack mechanisms across the range of space and ground-based systems (Haines, 2021, 8, 11; La Rocca, 2022, 29).

Nearly every KE ASAT results in a large amount of space debris, which causes an abundance of future and immediate problems for space activities, including endangerment of the basic military and commercial functions of satellites for the Global Positioning System (GPS), communications, and recreation. Space debris is therefore a highly undesirable side effect for any nation to risk and potentially dangerous to the integrity of a nation's armed forces. David Koplow (2014) addresses this issue in a substantially relevant and logical way in his article "An Inference about Interference: A Surprising Application of Existing International Law to Inhibit Anti-Satellite Weapons." His stated thesis is as follows: "The [National Technical Means] NTM-protection provisions of arms control treaties already prohibit the testing and use of destructive, debris-creating ASATs, because it is foreseeable that the resulting cloud of space junk will, sooner or later, impermissibly interfere with the operation of another state's NTM satellite, such as by colliding with it or causing it to maneuver away from its preferred orbital parameters into a safer, but less useful, location" (Koplow, 738-739). By "interfering" with these NTM verifications mandated by multiple treaties, Koplow suggests that intentional actions creating space debris are already outlawed by international law, and that the development of debris-creating KE ASATs should cease and be banned immediately (Koplow, 738).

Laser weapons, particle beams, and weapons containing depleted uranium are also under debate due to their radioactivity, as well as nuclear processes used for some of their operations. Some posit that nuclear activities or materials within a weapon system should constitute classifying them as nuclear weapons, thereby outlawing them in space per the Outer Space Treaty's nuclear weapons ban (Crockett, 2012, 687–688). Advocates for these weapons declare that the weapons are not nuclear. Of the three primary types debated, laser weapons use a nuclear or chemical reaction process to fire a radioactive beam, particle beams rapidly fire atomic charged particles at a target, and hypervelocity rod bundle weapons and railguns use depleted uranium as ammunition (Crockett, 2012, 674–682). Finally, the potential exists for the use of a nuclear explosion in space designed to generate an electromagnetic pulse (EMP) attack on an Earth target, which the RF "has worked on developing" in the form of an "EMP ASAT" (Crocket, 2012, 680). The RF and PRC are aggressively pursuing ASAT weapon advancements and preparing for space combat operations, including the RF recently fielding a ground-based laser weapon even as it publicly advocated for space not to be weaponized (Coats, 2019, 17). With the RF's recent developments in ASATs and its stated intent "to station weapons in space" (Clapper, 2016, 9–10), the complete weaponization of space by the RF and other nations—including the U.S. and the PRC—is inevitable, which leads to the question of which nation and which ideology will govern mankind's expansion into the stars.

5. The Future of Space

Space exploration converges on two of Sun Tzu's concepts of the strategic battlespace: "open ground" and the "ground of intersecting highways." The former consists of areas where all sides have "liberty of movement" and the latter of areas where "contiguous states" converge (Tzu, 1910, 46–47). On open ground, Sun Tzu advises not "to block the enemy's way," and on intersecting grounds he suggests to "join hands with your allies" (Tzu, 1910, 47). Space is essentially a combination of these types of ground, where all nations are contiguously connected, and yet it consists of a legally recognized area of free movement for all persons and nations. Interestingly, Sun Tzu's *The Art of War*, written over 2,000 years ago, advocates indirectly for peaceful human expansion into space, where allied nations proceed forth together while intentionally avoiding negative engagements with potential adversaries. Unfortunately, the PRC appears not to be adhering to this wise Chinese philosophical concept, and is instead positioning itself to ensure PRC domination and authoritarianism in the space domain. Interestingly, Sun Tzu's ancient concept of human cooperation and peaceful coexistence is more consistent with the U.S. Department of Defense's (DOD) and intelligence community's (IC) *National Security Space Policy* and the *National Space Policy of the United States of America*, than with the PRC government's policies (Department of Defense, 2011; Obama, 2010).

Executive Order (EO) 13914, signed on 6 April 2020, clarifies the position of the U.S. government that while international cooperation in space exploration is essentially mandatory, America "does not view [space] as a global commons," reiterating that the Outer Space Treaty does in fact protect the individual interests of nations in space, including the right to self-defense (Executive Order, 2020, 1). The policy further clarifies the intent of the United States to harvest materials from celestial bodies and strengthens the implied relationships with both the international community and the private sector concerning space exploration and related developments (Executive Order, 2020, 1). By combining these principles, this renewed position on space developments further complements Sun Tzu's ideas of the strategic battlespace in relation to the space domain moving into the future, regarding space as an area that can be used and exploited by everyone, but acknowledging that claims, defense, and security are also going to be an essential factor in the way mankind moves forward in the space domain.

In addressing the impact of space exploration, and the subsequent superiority gained by the PRC, the RF, or the U.S. in the process, it is important to recognize the three principle issues of the strategic space environment outlined in U.S. national policies: congestion, contestation, and competitiveness. The U.S. IC is mandated by section 1.1 of EO 12333 to "provide ... the necessary information on which to base decisions concerning the development and conduct of foreign, defense, and economic policies, and the protection of United States national interests from foreign security threats," which now include threats from space and threats toward U.S. space assets (Executive Order, 2008, 1). Congestion, contestation, and competitiveness in space now directly impact the IC's ability to effectively pursue its mandate under EO 12333 and must be addressed collectively to ensure the future national security of the United States on Earth and in space. Enhancing the space industrial base's ability to innovate and participate in the expansion of humankind into space fosters a unique opportunity to share with, and benefit from, research and development initiatives related to activities in space.

Combining private sector and government resources together has the potential to greatly accelerate advancements across a wide range of space assets—including spacecraft developments, zero gravity research, energy production, and weapon applications—all of which will help minimize the risks of congestion, contestation, and competitiveness. Congestion in space refers to objects, including active devices and dangerous debris, filling up the usable orbital paths used for government and commercial purposes, primarily satellites. It also applies to finite amounts of bandwidth and frequencies used for transmissions that are currently being exhausted by demand threatening to exceed supply (Department of Defense, 2011). Quantum communication technology research is advancing, with benefits that include unbreakable encryption, and

"unhackable satellite services," resulting in "secure communications and signals" transmissions that are "impossible to eavesdrop" using Quantum key distribution, which China successfully achieved in a test in 2020 at a distance over 1,000km (NATO, 2020, 73; La Rocca, 2022, 83, 32). Quantum communications developments also have the potential to decrease congestion of bandwidth and frequencies used for current transmissions, as they operate outside the radio frequency spectrum.

Developments in unmanned vehicle technologies, including swarming technology paired with artificial intelligence, offers a potential solution to space debris (NATO, 2020, 59–66). Swarms of miniaturized space-capable unmanned vehicles with high-powered laser technologies could be deployed to target and eliminate space debris to reduce congestion of near-Earth space. Congestion will also inherently refer to space traffic once an industry exists that requires transportation between the Earth and the Moon, as well as to physical locations for lunar and Martian resource exploitation facilities, extraction points, and places to build and operate on celestial bodies, including the Moon and Mars. This will eventually include a significant focus on the colonization of Mars since large portions of the planet are unsuitable for human habitation due to terrain, radiation, meteoroids, and weather. Short-term intelligence and counterintelligence impacts from the congestion of near-Earth space consist of primarily radio interference, protecting satellites from becoming compromised, effective deployment and concealment of collection platforms, and ensuring the integrity of protected information in transit.

Sharing space in accordance with Sun Tzu's ancient wisdom does not mean ceding it, and while space debris is the primary factor in congestion, contestation is becoming an issue due to potential adversarial ASATs. Contestation is an anticipated inevitability that will grow exponentially as more nations enter space and with further developments and potential use of ASATs, either in war, by accident, or for other reasons. Murphy's Law applies, especially in space. Currently, competitiveness is driving both the potential for contestation as well as the congestion in near-Earth space. Commercial and multi-governmental competition is increasing for space-related research and development, deployment of assets, and physical space for occupation by those assets. Intelligence agencies in many nations, including allies and adversaries of the U.S., are now advancing the deployment, use, and decision advantages of spaceborne intelligence assets, including space-based surveillance and weapons platforms. Reasserting U.S. superiority over the space environment is vital to the continuation of American leadership on Earth and the effectiveness of IC assurance of national security through space superiority. American leadership in space exploration is the only way to ensure that humanity's expansion into the stars is undertaken with the ideologies of liberty and free-market economics leading the way.

America's leadership in ingenuity and technological developments, combined with free-market capitalism, has transformed the face of the world for more than two centuries. Its leadership has created the environment necessary to explore game-changing space technologies, many of which will revolutionize the entire space industry. For example, the Variable Specific Impulse Magnetoplasma Rocket (VASIMR) is an experimental electromagnetic thruster for spacecraft propulsion that will dramatically reduce travel time to Mars and other destinations (Krishna and Kumar, 2014). Commercial spacecraft like the Dream Chaser Cargo System will result in a private sector space travel industry, incentivizing space tourism and, potentially, a space cargo transportation industry (Gold, 2016, 1). SpaceX has begun launching its *Starlink* communication satellite constellation to provide global connectivity, and as of February of 2021, *Starlink* already contained more satellites in orbit than the PRC, with plans to have 12,000 satellites in orbit by 2027 to complete the system (NATO, 2020, 81; Harrison, 2021, 2).

In February 2020, the U.S. department of Energy announced a $50 million investment in Fusion research and development projects across the country (Department of Energy, 2020). One of these is the Plasma Science and Fusion Center at the Massachusetts Institute of Technology with the goal of keeping the United States at the forefront of fusion energy development (Rivenberg, 2020). Another is the Fusion Technology Institute at the University of Wisconsin, which is focusing on advancing research in the field of helium-based fusion power production technologies on Earth (Dobransky, 2013). This technology will address finite terrestrial energy resources and production of ^{3}He-based electricity from lunar regolith. These are just a few examples of the future of space technology research and development, and such technologies were all made possible because of the structure of the American free-market system. The only way to prevent authoritarian leadership in the space domain is to provide an alternative, with liberty and free-market economics driving expansion into space.

Conclusion

The Artemis program concept has the potential to become a global space exploration initiative that benefits all life on Earth, creating opportunities for advancements across the entire spectrum of human life and well-being. The possibility of fusion power production will dramatically impact Earth's energy industry, off-setting the economic balance of power for generations. Ideological power struggles on Earth will inevitably bleed into the space domain impacting how humans are governed in space and on celestial bodies, dictating whether or not freedom and democracy survive. As technologies shrink the world, they are also shrinking space, creating ease of access and commercial opportunities that have never been possible throughout mankind's history. The international community will eventually be forced

to unite as one Earth or fall as a house divided, and the implications of space developments are accelerating this decision. Instead of focusing on how best to 'win' in the areas of congestion, contestation, and competitiveness in space, nations should focus on the best ways to reduce these issues, uniting to eliminate space debris, to cooperate in clean fusion energy development, to commercialize space operations, and to lead the world forward into a new era.

The PRC's and RF's posture in and towards space are driving the world towards conflict, and America is the only nation positioned to counter their actions. Space opportunities will inevitably result in new ways and means to achieve power and control, and if the U.S. does not achieve both, then the PRC or the RF will. The biggest challenge for America and the IC will be to balance President Dwight Eisenhower's vision with Sun Tzu's battlefield strategies. Eisenhower understood in 1958 that "through [space] exploration, man hopes to broaden his horizons, add to his knowledge, and improve his way of living on earth" (Office of the Historian, 1958, 2). Sun Tzu knew that "all warfare is based on deception," "the highest form of generalship is to balk the enemy's plans," and the greatest fighters "put themselves beyond the possibility of defeat" to achieve victory (Tzu, 1910, 3, 8, 12). American leaders participating in seizing and maintaining U.S. space superiority shoulder this responsibility and must forge a new path forward that enhances human life on Earth, denies the possibility of victory to U.S. adversaries, and ensures the in-

tegrity and security of American assets
in the space domain as the world moves
forward together into the stars.

References

Calcagno, E. (2022). NATO and its Members: A Space Alliance? in The Expanding Nexus Between Space and Defence. *Istituto Affari Internazionali 22*(1): 35–43. https://www.iai.it/sites/default/files/iai2201.pdf.

Chanock, A. (2013). "The Problems and Potential Solutions Related to the Emergence of Space Weapons in the 21ˢᵗ Century." *Journal of Air Law and Commerce 78*: 697–698.

Clapper, J. (2016). "Worldwide Threat Assessment of the U.S. Intelligence Community." *Office of the Director of National Intelligence*. 9 February. https://www.dni.gov/files/documents/SASC_Unclassified_2016_ATA_SFR_FINAL.pdf.

Coats, D. (2019). Worldwide Threat Assessment of the U.S. Intelligence Community. *Office of the Director of National Intelligence*. 29 January. https://www.dni.gov/files/ODNI/documents/2019-ATA-SFR---SSCI.pdf.

Crawford, I. (2015). Lunar Resources: A Review. *Progress in Physical Geography 39*(2): 137–67.

Crockett, J. (2012). Space Warfare in the Here and Now: The Rules of Engagement for U.S. Weaponized Satellites in the Current Legal Space Regime. *Journal of Air Law and Commerce 77*: 687–88.

Defense Intelligence Agency. (2022). Challenges to Security in Space: Space Reliance in an Era of Competition and Expansion. *Defense Intelligence Agency*. https://www.dia.mil/Portals/110/Documents/News/Military_Power_Publications/Challenges_Security_Space_2022.pdf.

Department of Defense. (2011). National Security Space Strategy: Unclassified Summary. *Department of Defense Archives*. January 3. https://www.dni.gov/files/documents/Newsroom/Reports%20and%20Pubs/2011_nationalsecurityspacestrategy.pdf.

Department of Energy. (2020). Department of Energy Announces $50 Million for Fusion Energy R&D. *Department of Energy*. 13 February. https://www.energy.gov/.

Dobransky, S. (2013). Helium-3: The Future of Energy Security. *International Journal on World Peace 30*(1): 61–88.

Executive Order 12333. (2008). United States Intelligence Activities of December 4, 1981, As Amended by Executive Orders 13284 (2003), 13355 (2004), and 13470 (2008). *Code of Federal Regulations.* Title 3. 30 July. https://dpcld.defense.gov/Port als/49/Documents/Civil/eo-12333-2008.pdf.

Executive Order 13914. (2020). Encouraging International Support for the Recovery and Use of Space Resources. *Federal Register 85*(70). 6 April. https://www.fed eralregister.gov/.

Garcia, M. (2021). Space Debris and Human Spacecraft. *NASA.gov.* Last Accessed 4 April 2022. http://www.nasa.gov/mission_pages/station/news/orbital_debris. html.

Gold, M. (2016). NASA at a Crossroads: Reasserting American Leadership in Space Exploration, Testimony, *Hearing of the Senate Space, Science, and Competitiveness Subcommittee.* 13 July. https://www.commerce.senate.gov/public/_cache/ files/22225dab-2f0d-437e-98d2-288c30ec9791/AC3538D71CDC78C585140 280AE91E184.mike-gold-testimony.pdf.

Haines, A. (2021). Annual Threat Assessment of the U.S. Intelligence Community. *Office of the Director of National Intelligence.* 9 April. https://www.odni.gov/files/ ODNI/documents/assessments/ATA-2021-Unclassified-Report.pdf.

Harrison, T., Johnson, K. & Young, M. (2021). Defense Against the Dark Arts in Space: Protecting Space Systems from Counterspace Weapons. *Center for Strategic and International Studies.* February. https://aerospace.csis.org/wp-content/uploads/ 2021/03/032321_HarrisonJohnsonYoung_DefenseAgainstDarkArtsInSpace_Re port_Update-compressed.pdf.

Hughes, J. (2011). Confusion Over Space. *The Journal of Social, Political, and Economic Studies 36*(1): 3–54.

Jensen, E. (2014). The Future of the Law of Armed Conflict: Ostriches, Butterflies, and Nanobots. *Michigan Journal of International Law 35*(2): 253–317.

Kolasinski, R. D., Shugard, A. D., Tewell, C. R. & Cowgill, D. F. (2010). Uranium for Hydrogen Storage Applications: A Materials Science Perspective. *Albuquerque, NM: Sandia National Laboratories.* August. https://www.osti.gov/servlets/purl/993 617.

Koplow, D. (2014). An Inference about Interference: A Surprising Application of Existing International Law to Inhibit Anti-Satellite Weapons. *University of Pennsylvania Journal of International Law 35*: 746–747.

Koplow, D. (2019). Deterrence as the MacGuffin: The Case for Arms Control in Outer Space. *Journal of National Security Law & Policy 10*(2): 293–349.

Krishna, V. C., & Kumar, A. S. (2014). Magneto Hydro Dynamics-Plasma Dynamic (MHD) for Power Generation and High-Speed Propulsion. *International Journal of Advances in Engineering & Technology 7*(1): 168–175.

La Rocca, G. (2022). Russia and China: West's Systemic Rivals on Orbits in The Expanding Nexus Between Space and Defence. *Istituto Affari Internazionali 22*(1): 25–34. https://www.iai.it/sites/default/files/iai2201.pdf.

La Rocca, G. (2022). The Technology Dimension and Duality in The Expanding Nexus Between Space and Defence. *Istituto Affari Internazionali 22*(1): 83–90. https://www.iai.it/sites/default/files/iai2201.pdf.

NASA. (2020). Artemis Plan: NASA's Lunar Exploration Plan Overview. *National Aeronautics and Space Administration.* https://www.nasa.gov/sites/default/files/atoms/files/artemis_plan-20200921.pdf.

NATO. (2020). Science & Technology Trends 2020–2040: Exploring the S&T Edge. *NATO Science & Technology Organization.* March. https://www.nato.int/nato_static_fl2014/assets/pdf/2020/4/pdf/190422-ST_Tech_Trends_Report_2020-2040.pdf.

Obama, B. (2010). *National Space Policy of the United States of America.* Washington, D.C.; The White House, 28 June. https://obamawhitehouse.archives.gov/sites/default/files/national_space_policy_6-28-10.pdf.

Office of the Historian. (1958). Statement of Preliminary U.S. Policy on Outer Space. *Department of State.* https://history.state.gov/historicaldocuments/frus1958-60v02/d442.

Pool, P. (2013). War of the Cyber World: The Law of Cyber Warfare. *The International Lawyer 47*(2): 299–323.

Rivenberg, P. (2020). Plasma Science and Fusion Center Receives $1.25M from ARPA-E to Explore Practical Paths to Fusion: MIT Experience with Hearing Plasmas will Support Novel and Low-Cost Approaches to Creating Fusion Energy. *Plasma Science and Fusion Center.* 15 April. https://news.mit.edu/2020/psfc-re

ceives-arpa-e-funding-to-explore-practical-fusion-paths-0415.

Tzu, S. (1910). *The Art of War.* Trans. Lionel Giles. London: Luzac and Co. https://sites.ualberta.ca/~enoch/Readings/The_Art_Of_War.pdf.

United Nations. (1967). Treaty on Principles Governing the Activities of States in the Exploration and Use of Outer Space, Including the Moon and Other Celestial Bodies, (hereafter referred to as the Outer Space Treaty). *U.S. Department of State.* 27 January. https://2009-2017.state.gov/t/isn/5181.htm.

United Nations. (1945). United Nations Charter (full text). *United Nations.* https://www.un.org/en/about-us/un-charter/full-text.

Space Education and Strategic Applications Journal • Vol. 3, No. 2 • Winter 2022

Biden Administration U.S. Space Force Policy Literature

Bert T. Chapman[a]

[a] Purdue University Libraries, 610 Purdue Mall, West Lafayette, IN, chapmanb@purdue.edu

Abstract

The U.S. Space Force (USSF) was established during the Trump Administration. With the 2021 transfer of power to the Biden Administration, there was some debate as to whether the newest U.S. armed service branch would continue in the new administration. Policy pronouncements from the Biden Administration and ongoing congressional oversight and appropriations demonstrate that USSF will remain a key part of the U.S. national security architecture. This work examines U.S. Government literature on USSF produced by the Biden Administration, other U.S. government agencies, and congressional oversight entities during the first three quarters of 2021. It demonstrates there is sufficient bipartisan support to financially and politically sustain USSF for the foreseeable future. How long this support is sustained remains to be determined based on managerial developments with USSF and the extent the international security environment places increasing emphasis on USSF using its evolving capabilities to address emerging threats to U.S. and allied national security requirements.

Keywords: United States Space Force, Biden Administration, U.S. congress, congressional oversight, China military space power, Russia military space power, space force education, and space force defense contracting

Literatura de política de la Fuerza Espacial de EE. UU. de la Administración Biden

Resumen

La Fuerza Espacial de los Estados Unidos (USSF) se estableció durante la Administración Trump. Con la transferencia de poder de 2021 a la Administración Biden, hubo cierto debate sobre si la nueva rama del servicio armado de EE. UU. continuaría en la nueva administración. Los pronunciamientos de política de la Adminis-

doi: 10.18278/sesa.4.1.4

tración Biden y la supervisión y asignaciones continuas del Congreso demuestran que la USSF seguirá siendo una parte clave de la arquitectura de seguridad nacional de EE. UU. Este trabajo examina la literatura del gobierno de los EE. UU. sobre el USSF producida por la Administración Biden, otras agencias gubernamentales de los EE. UU. y entidades de supervisión del Congreso durante los primeros tres trimestres de 2021. Demuestra que existe suficiente apoyo bipartidista para sostener financiera y políticamente al USSF en el futuro previsible. Queda por determinar cuánto tiempo se sostendrá este apoyo en función de los desarrollos gerenciales con USSF y la medida en que el entorno de seguridad internacional pone un énfasis cada vez mayor en USSF utilizando sus capacidades en evolución para abordar las amenazas emergentes a los requisitos de seguridad nacional de EE. UU. y aliados.

Palabras clave: Fuerza Espacial de los Estados Unidos, Administración Biden, Congreso de los Estados Unidos, Supervisión del Congreso, Poder espacial militar de China, Poder espacial militar de Rusia, Educación de la fuerza espacial y Contratación de defensa de la fuerza espacial

拜登政府下的美国太空军政策文献

摘要

美国太空军（USSF）成立于特朗普政府时期。随着 2021 年权力移交给拜登政府，关于这一最新的美国武装部队是否会在新政府中继续存在一事出现了争论。拜登政府的政策声明以及正在进行的国会监督与拨款表明，USSF 仍将是美国国家安全架构的关键部分。本文分析了2021 年前三个季度中拜登政府、其他美国政府机构以及国会监督实体所制定的关于 USSF 的政府文献。分析表明，在可预见的未来，两党会提供充足的支持，以在财政和政治上维持 USSF。这种支持的持续时间仍取决于 USSF 的管理发展以及国际安全环境在多大程度上越来越重视 USSF，使用其不断发展的能力来应对美国与盟国国家安全要求所面临的新威胁。
关键词：美国太空军，拜登政府，美国国会，国会监督，中国太空军事力量，俄罗斯太空军事力量，太空军教育，太空军国防承包

Introduction

Legally established during the Trump Administration by the Fiscal Year 2020 National Defense Authorization Act on December 1, 2020 (Public Law 116-92), the United States Space Force (USSF) has overcome temporarily brief hyper-partisan opposition from some elements of the Biden Administration's political base to become a regular part of the U.S. military as documented in U.S. Government and military information resources. Groups such as Greenpeace, Code Pink, the Union of Concerned Scientists, Win Without War, and Physicians for Social Responsibility argued:

> The proposed Space Force will create an unnecessary bureaucracy that will cost taxpayers over $16 billion in fiscal year 2021 alone, and tens of billions more in the coming years, while focusing U.S. efforts on militarization rather than cooperation in space, increasing the risks to U.S. military and civilian space assets. (Wolfgang, 2020).

Hopes for USSF abolition were destroyed early in the Biden Administration. On February 2, 2021, White House press secretary Jen Psaki announced: "We look forward to continuing the work of the Space Force and invite the members of the team to come visit us in the briefing room and share an update on their important work." USSF's continuance was also demonstrated by its strong support within the military and bipartisan congressional support including March 2, 2021 announcements by House and Senate Armed Services Committee chair Representative Adam Smith (D-WA) and Senator Jack Reed (D-RI), with Smith spokesperson Monica Matoush and Reed spokesperson Chip Unruh publicly verifying their support for USSF as a standalone service. Congressional Republican support for USSF was by Representative Mo Brooks (R-AL), who said "I will fight any effort to minimize or eliminate the Space Force as a separate branch of America's military" (Halaschak, 2021; Perano, 2021).

Early Biden Administration Policymaker Statements

Early indication of possible Biden Administration thinking on USSF was reflected in answers to Advance Policy Questions (APQ) submitted by the Senate Armed Services Committee to eventual Secretary of Defense Lloyd Austin, Deputy Secretary of Defense Kathleen Hicks, and Undersecretary of Defense for Policy Colin Kahl as part of their confirmation process. During his January 19 confirmation hearing, Austin did not refer to USSF in his opening statement (Austin, 2021). Responding to the committee's question about whether USSF's creation was warranted and whether he recommended changes in its structure, authorities, or missions, Austin contended that creating USSF reflected recommendations and advice from multiple independent commissions, Congress, and multiple presidential administrations while acknowledging establishing USSF as the armed forces sixth branch represented significant organizational changes within the Defense Department (DOD). He also stressed that DOD's space enterprise was not well-integrated with service and other terrestrial commands.

He went on to note that the 2020 *Defense Space Strategy* provides substantive detail on space and counterspace threats posed by China and Russia and the importance of working with allies and partners to ensure unfettered access and freedom for operating in space. Responding to a question about whether he supported developing offensive and defensive space systems to counter threats in the space warfighting domain, Austin contended: "A balance of offensive and defensive capabilities, as well as resilient architectures, are essential to any credible strategy to deter hostile action and protect vital U.S. interests should conflict extend to space" (U.S. Congress, Senate Armed Services Committee, 2021(a)).

Nascent Biden Administration USSF stances were further reflected in APQ's Hicks answered on February 2, 2011. Hicks asserted that the Trump Administration's 2018 *National Defense Strategy* (NDS) presented a broadly accurate assessment of the increasingly complex and volatile national security environment produced by China and Russia. She went on to assert that growing Chinese and Russian counterspace arsenals are the most immediate security threats to U.S. and allied space activities, and emphasized that Chinese and Russian military space doctrines consider space as crit-

ical to modern warfare, reducing military effectiveness, and winning future wars. Hicks went on to stress that the U.S. should not expect its adversaries to discriminate between commercial and military satellites in peacetime competition or military conflict; that the U.S.' technological edge is continually eroding by aggressive and on-going efforts by foreign nations to illicitly gather and adopt advanced proprietary system designs and processes the U.S. space industrial base. She also concluded by stressing that she is open to the National Reconnaissance Office (NRO) needing to work with U.S. Space Command to integrate and synchronize operations (U.S. Congress, Senate Armed Services Committee, 2021(b)).

During his March 4, 2021 response to a Senate Armed Service Committee hearing (APQ), Kahl announced he would assess (DOD) readiness to implement NDS and the 2020 *Defense Space Strategy* by determining personnel readiness and space-based system resilience to address current and emerging challenges. Noting the leading counterspace threats posed by China and Russia, Kahl also noted Iran and North Korea possess some counterspace capabilities capable of threatening U.S. and allied satellites. Kahl proceeded to maintain that developing both offensive and defensive space capabilities is essential for an effective U.S. strategy for deterring and countering hostile use of space along with freedom of operation in, from, and to this domain. He went on to stress he would review whether existing armed service space capabilities should be retained subsequent to

USSF's creation; is open to working to determine how commercial technology in launch and space applications can be used for mission assurance and warfighting requirements; and favored the recommendation contained in the FY 2020 *National Defense Authorization Act* to select a nominee as Assistant Secretary of Defense for Space Policy (U.S. Congress, Senate Armed Services Committee, 2021(c)).

March-May 2021 Developments

In March 2021 the Biden Administration released its *Interim National Security Strategic Guidance*. This document can be viewed as a potential forerunner to the congressionally-mandated *National Security Strategy of the United States*, which was published in October 2022. References to USSF in *Interim Strategic Guidance* were very cryptic and confined to this sentence: "We will explore and use outer space to the benefit of humanity, and ensure the safety, stability, and security of outer space activities" (Biden, J. R., 2021).

Further explanation of emerging Biden Administration USSF policy was provided during March 13, 2021 congressional testimony. Acting Under Secretary of Defense for Research and Engineering Barbara McQuiston told the Senate Appropriations Committee's Defense Subcommittee that DOD was investing in technologies and studying capabilities to defeat regional hypersonic weapons with a key first element of this involving detecting and tracking incoming missile threats. She mentioned that the Missile Defense Agency (MDA) has delivered a real-time sensing and hypersonic vehicle tracks for the Indo-Pacific Region with this capability being achieved in collaboration with industry partner and other U.S. geographic military and functional commands. This capability will be a critical component in hypersonic missile defense by providing a persistent and layered capability to detect and track dim boosting ballistic missiles, hypersonic glide vehicles, and raids in all flight phases. MDA has made two awards to industry to build an on-orbit prototype space vehicle for a planned 2023 launch while DOD's Space Development Agency (SDA) will transition to USSF during Fiscal Year 2023 as this armed service's acquisition agency (U.S. Congress, Senate Committee on Appropriations, Subcommittee on Defense, 2021).

Recommendations for USSF deployment have been produced by multiple think-tanks representing divergent policy perspectives. In May 2021 the conservative Center for Security Policy made the following five recommendations for the Biden Administration:

- Approach the issue of space weapons with the understanding that the space weapons debate is a straw man argument that implicates space control and involves legal, political, and psychological aspects.

- Before enacting policy to address space security take into consideration that the concept of Western deterrence is not comparable to deterrence recognized by geopolitical rivals, which does not preclude the use of counter-space capabilities.

- Take into account the dual-use nature of space technology creates the potential for gray zone operations in outer space that could form customary international law/norms unfavorable to U.S. interests.

- Proactively support the use of innovative outer space technologies and activities by non-governmental operators and resist attempts by geopolitical competitors to conflate these activities with space weapons.

- Exercise caution towards any proposed agreements banning or limiting the use of so-called space weapons that could be ignored in the event of hostilities (Listner, 2021).

On May 20, 2021, the Senate Armed Services Committee held a hearing on U.S. Strategic and Space Command's FY 2022 congressional budget request and its implications for USSF. Space Command Commander General John Dickinson testified that space is a warfighting domain and that both China and Russia have expanded their capabilities from direct-assent Anti-Satellite (ASAT) weaponry and direct energy efforts in electronic warfare arenas including jamming and laser technology. He stressed that the U.S. military space assets are well integrated in space and cyber operations and responded to a question from Senator Mike Rounds (R-SD) on whether the U.S. had the ability to defend against new threats and maintain unrestricted access to space by stressing the importance of having satellites and terrestrial assets to understand Chinese and Russian orbital activities. Dickinson also told Senator Rick Scott (R-FL) his belief that the U.S. was on a "glide path" to being able to protect and defend orbital capabilities and capacities (U.S. Congress, Senate Armed Services Committee, 2021(d)).

Later, in May 2020, the USSF released *U.S. Space Force Vision for a Digital Space Service*. This vision document stressed that becoming a digital service was a warfighting imperative for USSF driven by the nature of the military threat and USSF's size. In this document's introduction, USSF Chief of Space Operations (CSO) General John W. Raymond argued: "We know potential adversaries are developing a spectrum of threats at an alarming pace, directly challenging stability in space and the many benefits we enjoy as a spacefaring nation. To counter these threats, we must change the paradigm. We must act far more swiftly and decisively across all aspects of leadership, acquisition, engineering, intelligence, and operations in order to take up permanent residence inside the adversary's observe, orient, decide, and act (OODA) loop. In addition, given the relatively small size of the USSF, accomplishing this goal will require us to amass a technologically adept, 'digitally fluent' space cadre more proficient, efficient, and agile than any other force in history" (U.S. Space Force, 2021(a)).

Figure 1. General John W. Raymond-Source: U.S. Department of Defense.

This document went on to stress that USSF must have three key tenets, including:

1) An interconnected force effectively and efficiently sharing relevant information with multiple stakeholders supporting the mission with data centricity and pervasive connective being hallmark characteristics of Guardians (as USSF personnel are called).

2) Innovation is a second USSF tenet

characterized by embracing new approaches and eagerly challenging the status quo stemming from a deliberate commitment to continuously evolve, improve, and adapt. This also involves equipping Guardians with correct toolsets to harness their skills and adaptively respond to adversary threats by being aggressive and early adopters of cutting-edge, user-driven technologies representing the best industry capabilities.

U.S. Space Force Vision for a Digital Service

SF/CTIO

MAY 2021

Figure 2. U.S. Space Force Vision for A Digital Service. Source: U.S. Space Force.

3) A third tenet is being digitally dominant by translating cumulative technical prowess into powerful force-multiplying effects for developing, fielding, and operating capabilities more quickly and quickly than potential adversaries. Achieving such lasting dominance requires instilling and synthesizing preexisting interconnected and innovative elements to all aspects of how USSF executes its mission of supporting the joint warfighting force (U.S. Space Force, 2021(a)).

Additional supplements of this vision include a digital workforce prioritizing data-centric solutions over product-centric processes; making optimum use of artificial intelligence routines or robotic process automation to free Guardians from allocating monotonous staffing activities and allow them to engage in training, educating, and wargaming to become a world-class fighting force; and using digital operations collectively to create a lethal space warfighting force ensuring digital dominance translating to space superiority maintenance. This document concludes with a scenario of how the USSF would respond to adversary counterspace threat scenarios with immediately identifying an emergent threat at T-0 minutes; establishing contingency responses at T+5 minutes including convening a virtual threat conference; initiating a virtual threat conference enabling real-time interaction within the data space and T+2 hours; Guardians being empowered to take collective action at T+12 hours; exploring and taking short-term countermeasures and long-term responses including updating software within T+36 hours; rapidly developing and testing digital solutions including high-fidelity emulations of target platforms within T+5 days; and placing enduring responses into place including ripple effects across the enterprise affecting concerning equipment, cost, schedule, and risk within T+6 to T+28 days (U.S. Space Force, 2021).

Business Contracting Developments

Business contracting can also produce high stakes legal and commercial implications for agencies aspiring to obtain lucrative government contracts from USSF and other government agencies. Inmarsat Government protested terms of Request for Proposals (RFP) to obtain a worldwide commercial broadband satellite services for the Navy's Military Sealift Command. It filed a protest against the Defense Information Systems Agency (DISA) with the Government Accountability Office (GAO) maintaining that DISA mailed to reduce competitive harm maintained by Inmarsat for inadvertently releasing its existing pricing for non-bandwidth commercial solution, and also protested that DISA failed to include Inmarsat's past performance as an evaluation factor for its contract application. GAO determined that DISA caused competitive harm by releasing Inmarsat's non-bandwith commercial pricing, but rejected their contention that had failed to prove that its past contract performance was an unreasonable factor in DISA's RFP (U.S. Government Accountability Office, 2021).

Opportunities for aspiring contractors to do business with USSF are provided by U.S. Federal Contractor (USFCR). Examples of grant opportunities and grant awards from USSF during the first three quarters of 2021 include DOD's Space Test Program, posted on August 26, 2021, providing access for science and technology experiments including military branches, interagency cooperation, universities, and international partners based in Albuquerque (U.S. Federal Contractor (USCFR), 2021(a)). On September 10, 2021, USSF's Space Warfighting Analysis Center (SWAC) announced a classified level business fair in Washington, D.C., on October 27, 2021, to communicate force design objectives to potential industry partners, which will be grounded in peer-review analysis using contemporary digital modeling tools and simulation environments (USCFR, 2021(b). A September 20, 2021 solicitation from Colorado's Peterson Space Force Base for ballistic and tempered glass panel fabrication and installation requirements is another example of a USSF contract opportunity (USCFR, 2021(c)).

According to an October 1, 2021 search of "space force" on USASpending.gov, there were 188 historic and ongoing contracts involving USSF awarded to companies as varied as Space Exploration Technologies, United Launch Services, Lockheed Martin, Silicon Mountain Technologies, Northrup Grumman, Parsons Government Services, and many others (USAspending.gov, 2021).

Biden Administration Congressional Budget Request and Congressional Reaction

In its May 2021 USSF congressional budget proposal, the Biden Administration requested the following totals for carry out USSF missions for Fiscal Year (FY 2022):

Operations and Maintenance	FY 2021 $2.569 billion	FY 2022 $3.441 billion
Procurement	FY 2021 $1.512 billion	FY 2022 $2.409 billion
Research, Development, Test, and Evaluation	FY 2021 $9.987 billion	FY 2022 $12.803 billion (U.S. Office of Management and Budget, 2021).

This submission began the annual congressional budget process, which over the following months saw congressional armed services and appropriations committees scrutinize and revise the administration's budget request and include additional provisions they require USSF to follow in the Fiscal Year 2022 *National Defense Authorization Act* (NDAA). This process sees witnesses from government agencies and non-government organizations invited to testify before congressional oversight committees and provide varying perspectives on funding and other expectations of government programs. This testimony and congressional questioning will result in appropriations bills for the Defense Department and other agencies which they are required to follow for federal fiscal years running from October 1-September 30 (McNellis, 2021).

During May 24, 2021 testimony before the House Armed Services Committee's Subcommittee on Strategic Forces Government GAO Director of Contracting and National Security Acquisitions Jon Ludwigson noted historic and ongoing DOD space weapons acquisition problems. Such difficulties have produced schedule delays exceeding five years, cost increases of hundreds of millions or billions of dollars, and program cancellations due to development problems. Ludwigson noted the DOD's 2019 establishment of a Space Development Agency (SDA) to unify and integrate department wide efforts to produce innovative satellite solutions, congressional and DOD efforts to increase the efficiency and deployment space capabilities, delays and cost increases in space acquisition programs as varied as the Space-Based Infrared System, Space Fence, Next Generation Overhead Persistent Radar, and insufficient information on investing in commercial cybersecurity technologies (Ludwigson, 2021).

Culminating challenges these extant problems will produce for the USSF include significant funding requirements for existing and emerging programs as the USSF assumes responsibility for existing defense space programs; the need to strike a right balance between new development methods and working within an existing knowledge-based acquisition framework to ensure cost, schedule, and performance goals are met (Ludwigson, 2021).

Testifying at this same hearing, USSF Vice-CSO General David Thompson noted that following the October 2020 establishment of Space Operations Command (SOC), two remaining field commands would be established during 2021 including Space Systems Command (SSC) with responsibility for developing, acquiring, and fielding operationally relevant capabilities in resilient and defendable architectures, and Space Training and Readiness Command (STARCOM) with responsibility for developing tactics, a testing enterprise, doctrine, and advanced operational training using warfighting professionals. SSC and STARCOM will be established after presidentially-nominated USSF officers are confirmed by the Senate. Thompson also announced planning had begun for a National Space Intelligence Center (NSIC) for providing foundational scientific and technical intelligence and operation-

al space intelligence to USSF, military combatant commands and the intelligence community along with (SWAC), currently with SOC leads analysis, modeling, wargaming, and experimentation to produce new operational concepts and USSF force design options while integrating these activities across DOD and the intelligence community (Thompson, 2021).

Testifying before the House Appropriations Committee's Defense Subcommittee on May 27, 2021, Joint Chiefs of Staff Chair General Mark Milley asserted:

> The Space Force investments accelerate modernization of the entire Joint Force. Space Force capabilities underwrite, enhance, and enable Joint Force operations. The Space Force protects U.S. capabilities and freedom of operation outpacing actions of our competitors. Ongoing actions to fully resource [USSF] include including transfer of the Space Development Agency (SDA) and unit transfers from the Army and Navy, will enhance the USSF's ability to organize, train, equip, and present forces who can compete, deter, and if necessary prevail should war initiate in, or extend to space. (U.S.Congress, House Committee on Appropriations, Defense Subcommittee, 2021)

An example of congressional support for the USSF was expressed by Rep. Mike Garcia (R-CA) during June 16, 2021 House floor debate. Stressing quickly evolving technological advances and geopolitical efforts by China and Russia to enhance their military space force capacities, Garcia stressed:

> …space is a domain where we, as a nation, can thrive, but it is also a domain where we can be vulnerable and susceptible to the malicious intent of foes such as China, Russia, and Iran. There are existential threats right now in space.
>
> Any decision on our part to divest from this adventure will not dissuade our foes from advancing their own space programs. It will only serve to highlight a massive strategic vulnerability and potentially create credibility gaps that will be impossible to fill in the future.
>
> A path of divestment is an unforgiving one and the damages irreparable. While we as a nation currently hold an advantage in space, our lead, like the nearly 118 years that separates today from the Wright Brother's first flight in 1903 can vanish in the blink of an eye. (Garcia, 2021)

Presenting the Air Force's posture statement during June 17, 2021 congressional testimony, Acting Air Force Secretary John Roth, Air Force Chief of Staff Charles Q. Brown, Jr., and USSF CSO Raymond noted that the U.S. is a space faring nation at its strongest when it has access to and freedom to operate in space. They went on to stress that potential adversaries are aware of this U.S. reliance on space and are engaging in efforts to reduce this freedom of operation. Examples of such Chinese and Russian measures include aggressively developing weapons to deny or destroy U.S. space capabilities in conflict; using mobile and ground-based laser and electronic warfare systems capable of jamming and blinding U.S. satellite systems; China investing in satellite grappling technologies such as the Shijian-17 satellite's robotic arm; and Russian testing of an on-orbit system which has released a projectile designed to destroy low-earth orbit U.S. satellites (Roth, Brown, and Raymond, 2021).

These policymakers went on to stress how the USSF's first year involved establishing a headquarters and Field Command structure aligning complimentary functions and streamlining command authority to pursue speed and agility. SOC the first field command, was stood up in October 2020 as the primary provider to the combatant commands. Two additional commands were installed during 2021 including Space Systems Command which will develop, acquire, and field operationally relevant and resilient space capabilities within resilient and defendable architectures, and Space Training and Readiness Command whose responsibilities involve developing tactics, a testing enterprise, doctrine, advanced warfare training, and a dedicated cadre of warfighting professionals. Space missions, billets, and financial resources have been transferred from 23 Air Force units to USSF, and preparations to merge operations, acquisition, and sustainment for some space systems distributed in existing military branches and the Office of the Secretary of Defense beginning in 2022. Additional USSF collaborations within DOD, interagency, commercial industry, and national allies is also occurring (Roth, Brown, and Raymond, 2021).

Efforts to increase digital capabilities, accelerating capability design, decision, and delivery are ongoing including missile warning and tracking capabilities; position, navigation, and timing; command and control including a Unified Data Laboratory which is a cloud-based, cyber-accredited, multi-classification data store facilitating universal data access in a data architecture partnership with the Air Force; and enhancing National Security Space Launch (NSSL) capabilities. Enhancements in training and doctrine are occurring through redesign across space operations competencies including elevating from basic space systems operation to threat and target-based advanced training and education. Areas encompassed within these advanced characteristics include: orbital warfare, space electromagnetic warfare, space battle management, space access and sustainment, military intelligence, cyber operations, and engineering and acquisition with ultimate aims of supporting tactics, strategies, and theories of victory (Roth, Brown, and Raymond, 2021).

On July 13, 2021, the House Appropriations Committee approved its version of the FY 2022 National Defense Authorization Act by a 33–23 vote. It recommended cutting funding from various Biden Administration USSF areas, including:

Operations and Maintenance

FY 2022 Budget Request	$3,440,712 billion
Committee Recommendation	$3,372,212 billion
Change from Budget Request	-$68,500,000

Procurement

FY 2022 Budget Request	$2,766,854 billion
Committee Recommendation	$2,741,708 billion
Change from Budget Request	-$25,146,000

Space Force, Research, Development, Test and Evaluation

FY 2022 Budget Request	$11,266,387 billion
Committee Recommendation	$10,774,318 billion
Change from Budget Request	-$492,069,000 (U.S. Congress, House Committee on Appropriations, 2021(b).

Congressional committee reports on agency funding requests also include written directions and reporting requirements for agencies to follow in executing congressional intent for their operations and programs. This is reflected in several sections of the House Appropriations Committee *FY 2022 NDAA* report. This committee expressed concern that the Air Force has not taken aggressive action to address space acquisition problems and made little progress in defining what the USSF will do differently than the Air Force. Additional concerns expressed by this committee's report include:

- Plans for the new Space Systems Command acquisition unit not resolving fundamental problems of overlap and role duplication, responsibilities, and authorities within existing Air Force space acquisition units.

- USSF needing a clear and concrete vision for future system architectures which are not philosophically based but grounded in rigorous technical analysis with executable plans and realistic budgets.

- DOD lacking a comprehensive system to measure the readiness and extent of satellite communications terminals, Global Positioning System receivers, and other terminals and user equipment have actually been fielded into platforms and systems designed to use such capabilities. The committee directed various DOD entities to prepare a space integration readiness report to congressional defense committees within 120 days after the 2022 NDAA is enacted.

- Expressed concern about the Next Generation Overhead Persistent Infrared Missile Warning Program having an unrealistic 2025 first geosynchronous satellite launch capability and emphasizing USSF providing unrealistic cost and schedule estimates undermines the credibility of its management of this and other programs.

- Continued supporting integrating commercial satellite communications capabilities into national security space communications architecture while wanting greater clarity from USSF on goals and strategy for its $23.4 million funding for this program and refusing to fund this program until the Air Force submits a report to congressional defense committees within 90 days of NDAA's 2022 enactment.

- Requiring DOD and the Director of National Intelligence to provide a plan to collect, consolidate, and characterize potential adversaries' laser threat activity data along with strategies to mitigate these threats within 120 days of NDAA's 2022 enactment.

- An earmark of $1.665 million to develop a core manipulator joint at the Texas A&M University Experiment Station inserted by Rep. Pete Sessions (R-TX) (U.S. Congress, House Committee on Appropriations, 2021(b)).

The House Armed Services Committee approved its version of the FY 2022 NDAA on September 10, 2021 on a 57–2 vote sending it for consideration to the full House (Key provisions of this committee's report included increasing USSF personnel from 6,434–8,400; increasing USSF's procurement budget from $2,766,854 billion to $2,773,54 billion; increasing the USSF's Research, Development, Testing, and Evaluation Budget from $11,266,387 billion to $11,594,787 billion; and increasing the USSF's Operations and Maintenance Budget from $3,440,712 billion to $3,751,412 billion (U.S. Congress. House Committee on Armed Services, 2021).

Provisions of this committee's report covering USSF include:

- Requiring the CSO to brief the House Armed Services Committee by January 31, 2022, on the effectiveness of USSF's higher education strategy in creating long-term strategic relationships, developing talent, and providing access to expertise, engineering, research, and development capability.

- Requiring USSF's Chief Technology and Innovation Officer to brief the House Armed Services Committee by December 31, 2021, on how USSF's Chief Information Officer will work with their DOD and Air Force counterparts to leverage cloud computing technologies for space programs.

- Requiring the U.S. Comptroller General to report to congressional defense committees by February 1, 2022, on U.S. space situational and domain awareness capabilities including analyzing the number and size of low-Earth orbit, geosynchronous earth orbit, and cislunar orbit tracked objects; review planned systems development and procurement of commercial space situational and domain awareness across the Future Years Defense Program including cost and schedule estimates; overview of the USSF's Unified Data Library including current volume, access to new observational data, U.S. Space Command usage; and recommendations to improve use of commercial space situational and domain awareness data services.

- Directing DOD to report to the House Armed Services Committee by February 1, 2022, on executing experimental spaceflight activities for next-generation launch vehicle systems and technologies relating to national security space launch applications pertinent to maintaining U.S. space technology superiority over China.

- A sense of Congress resolution requiring CSO to report to congressional defense committees on the USSF's "Range of the Future," identifying legal authorities which must be changed to address long-term infrastructure challenges to physical launch ranges and proposals to enhance infrastructure improvements at these sites including congressional action to implement these proposals.

- Requiring the CSO to report to congressional defense committees by February 25, 2022, on the most likely and dangerous threats to American space dominance within the next three to ten years, options to maintain this U.S. dominance, and actions required to support such dominance. Emphasis on maintaining U.S. freedom of movement on the Moon and in lunar and cislunar space was also stressed by the committee (U.S. Congress, House Committee on Armed Services, 2021).

Despite wide levels of support for the USSF in the House, some segments of opposition remain. This was demonstrated by the September 22, 2021 introduction of H.R. 5335, the *No Militarization of Space Act*, which proposed abolishing the USSF. Introduced by Rep. Jared Huffman (D-CA) and four other House members, this legislation was referred to the House Armed Services Committee where it is likely to die given current significant bipartisan support for the USSF (*H.R. 5335*, 2021).

On September 22, 2021, the Senate Armed Services Committee approved its version of the 2022 NDAA. It directed USSF to spend an additional $3 million on battery life cycle improvements and an additional $5 million increase to develop microelectronics to withstand space radiation. This committee also maintained the House Committee's FY 2022 USSF personnel authorization at 8,400, while recommending legislative language giving the Secretary of the Air Force greater authority to establish USSF personnel levels than permitted for other armed service branches (U.S. Congress, Senate Committee on Armed Services, 2021(e)).

Additional Senate USSF provisions for FY 2022 NDAA include:

- Directing the Chief of Space Operations to brief congressional defense committees by February 28, 2022, on the possibility of using prize authority authorized under 10 *USC* 2374(a) for launch responsiveness to replace key national security satellites during a conflict with particular emphasis on pushing launch capabilities to technological limits in payload size, payload numbers, and launch sites.

- Encouraging USSF to consider developing and deploying small flexible communication satellites capable of meeting connectivity and reconstitution objectives.

- Directing DOD, the Chairman of the Joint Chiefs of Staff, and USSF to give a detailed briefing to congressional defense committees by March 31, 2022, on the commercial space-based Intelligence, Surveillance, and Reconnaissance (ISR) needs of combatant commands through varying weather conditions. This briefing should include descriptions of existing space-based ISR combatant commands and a subset of requirements that can be met with commercial assets; analysis of how DOD is or will be leveraging commercial space-based solutions to meet combatant commanders requirements in the next five years; determining DOD's strategy to address these needs for the purchase of commercial satellite communications; assessing risks of over-reliance on commercial space-based ISR during conflict in space and other domains; assessing combatant commands ability to directly task space-based ISR for individual interest areas; and determining whether military space leaders can economically and competitively make bulk purchases of commercial space-based ISR for combatant commands comparable to bulk purchases of commercial satellite communications.

- DOD and other applicable executive branch entities reporting to congressional defense committees by April 30, 2022, to ensure adequate and timely communications occur between the U.S., China, and Russia on avoiding space debris collisions (U.S. Congress, Senate Committee on Armed Services, 2021(e)).

The Senate Committee authorized $768 billion in defense spending with USSF procurement of $2,799,354 billion; $3,751,912 billion for USSF operation and maintenance, and $11,795,166 billion for USSF research, development, testing, and evaluation (U.S. Congress, Senate Committee on Armed Services, 2021(e)).

Continuing congressional debate on the USSF will occur in floor debate and in committee deliberations. The September 22, 2021 House debate language mentions USSF seven times (*Congressional Record*, 2021(a)). On September 23, 2021, the House approved the FY 2022 NDAA with a bipartisan vote of 316–113 increasing defense authorizing spending $23 billion from the Biden Administration's request of $741 billion, and increasing overall USSF spending to $18,119,353 billion representing a 3.53% increase over the Administration's initial budget request (*Congressional Record*, 2021(b)). Differences between House and Senate versions of the FY 2022 NDAA will need to be resolved and may take several weeks or months before the final version of this legislation is submitted to and receives presidential approval. Evidence for this protracted period is provided by the FY 2021 NDAA, which was not signed until December 21, 2020, as part of a consolidated multiagency appropriations bill (Public Law 116-260). FY 2022 USSF space funding was not finalized until Public Law 117-103 was signed on March 15, 2022, with operations and maintenance receiving $3,435,212 billion; procurement receiving $3,023,408 billion; and research and development receiving $11,597,405 billion for a cumulative total of $18,455,625 billion (Public Law 117-103).

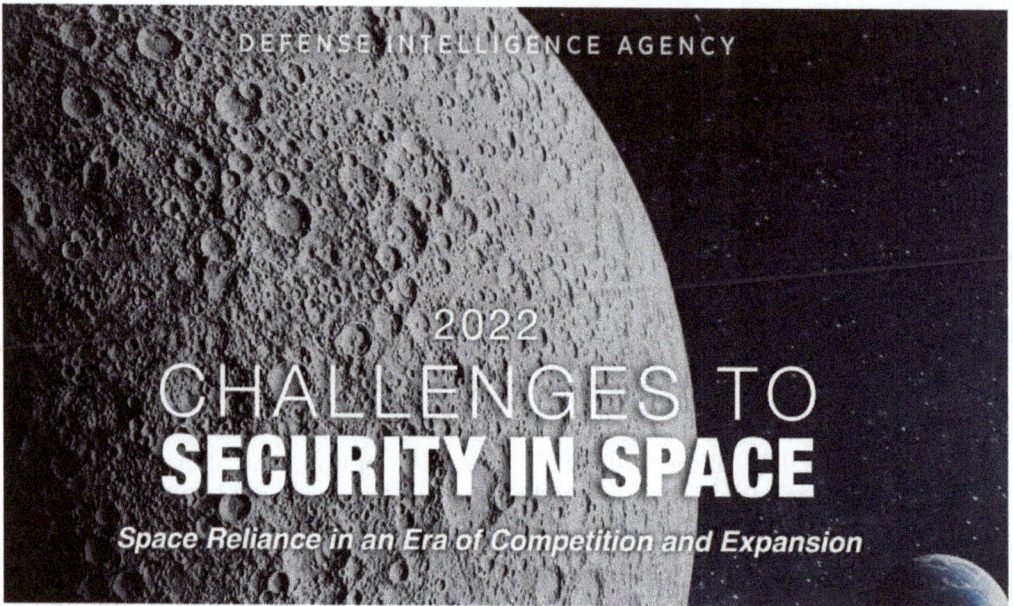

2022 Defense Intelligence Agency Report

The trajectory of USSF funding and missions will be acutely influenced by the military space developments of other countries. These have been documented in a 2022 Defense Intelligence Agency (DIA) report on emerging space security challenges confronting the U.S. and its allies. This assessment noted that between 2019–2021, the number of Chinese and Russian satellites in orbit had increased by nearly 70% with these satellites including those covering satellite communications, remote sensing, navigation-related, and science and technology development and other purposes including geolocation and tracking of friendly and adversary activities, target identification, and the ability to conceal sensitive tests, evaluation activities, and military exercises and operations (U.S. Defense Intelligence Agency, 2022).

DIA noted that China is likely developing weapons to use against orbiting satellites to degrade and deny adversary space capabilities; considers space superiority as critical to conducting informationized warfare; actively seeks overt and covert acquisition of foreign space and counterspace technologies including through exploitation of overseas scholars; and increased its satellite intelligence, surveillance, and reconnaissance fleet to over 250 systems, making it second to the U.S. as of January 2022. Russian military doctrine also views space as a warfighting domain and believes achieving space supremacy is a decisive factor in winning future wars. While rhetorically promoting international space arms agreements, DIA also asserts that Russia is developing counterspace capabilities to attack U.S. and allied space assets; created an Aerospace Forces military branch in 2015; sees space as an American Achilles heel while striving to develop counterspace systems to neutralize U.S. space-based military and commercial systems to offset perceived U.S. advantages in these areas; Moscow actively seeks to acquire foreign space and counterspace technologies despite international sanctions; and seeks to use Directed Energy Weapons to blind satellite sensors. DIA also assesses that Iran and North Korea are also seeking to develop space and counterspace capabilities to enhance its security objectives and interfere with U.S. space operations (U.S. Defense Intelligence Agency, 2022).

University Partnerships

USSF has sought to increase its impact by collaborating with universities with strong science and technology backgrounds and reputations. USSF's University Partnership Program strives to accomplish the following:

- Establish opportunities for world-class research, advanced degrees, and workforce and leadership development.

- Identify and pursue research areas of mutual interest.

- Establish scholarship, internship, and mentorship opportunities for university students and Reserve Officer Training Corps (ROTC) cadets.

- Recruit and develop multifaceted officer, enlisted, and civilian Guardians emphasizing Science, Technology, Engineering, and Mathematics.

Universities signing Memorandums of Understanding (MOU) with USSF during FY 2021 include: Georgia Institute of Technology, Howard University, Massachusetts Institute of Technology North Carolina Agricultural and Technical State University, Purdue University, University of Colorado System (beginning with Boulder and Colorado Springs), University of North Dakota, University of Texas System (beginning with Austin and El Paso), and University of Southern California (U.S. Space Force, 2021(b)).

Conclusion

USSF receives continuing support from Congress and the Biden Administration which continues from its establishment during the Trump Administration. Its ongoing support is testimony to increasing recognition of the growing importance of space as an arena of international security competition, which has surpassed the partisan and ideological differences between the Trump and Biden Administrations. The USSF is also responsible for guiding and working with the commercial space industry, piloting orbital test vehicles, training for orbital warfare, and remaining aware of and competitive with Chinese and Russian space activities (Weires, 2021). Army and Navy satellite communication (SATCOM) missions have now shifted to Space Force. USSF personnel are already involved and will continue being involved in contributing written analyses of their service programs (Garamone, 2021; Poole, Bettinger, and Reith, 2021).

USSF, DOD, and congressional oversight entities, and other entities will continually scrutinize military space programs and their collaborations with the commercial space industry. Addressing managerial dispersion and acquisition effectively will be key USSF challenges. Topics such as national security space launch, the electromagnetic spectrum, remote sensing, Chinese, Russian, and other national responses to USSF activities will keep DOD, USSF, and congressional overseers and appropriators busy in the next several years and produce voluminous documentation with significant quantities of this being publicly accessible for U.S. taxpayers to evaluate (McCall, 2020; McCall, 2021; Hoehn, Gallagher, Saylor, 2021; Townsend, 2021; Bjørkum, 2021).

Despite its young age, USSF contracts have been dispersed to companies in states as varied as California, Colorado, Florida, Maryland, Nebraska, New Jersey, Ohio, and Virginia. This is likely to continue increasing and give these areas congressional Representatives and Senators additional motivation to enhance their re-election prospects by ensuring these contracts are sustained for their constituents employed in these businesses and for their overall economic impact although the significance of defense spending on electoral prospects and economic impact is disputed (USASpending, 2021; Thadalikit, 2001; Barro & de Rugy, 2013).

USSF receives continuing support from Congress and Biden Administration. It remains to be seen how long this support will last as the U.S. and its allies strive to recover from the COVID-19 pandemic and respond to increasing Chinese and Russian aerospace threats in light of the debacle of Afghanistan. Russian military aggression against Ukraine will likely increase U.S. financial and political support for USSF. Public support for USSF will also depend on program performance and progress in subsequent years and whether concern over the continually upward spiraling national debt and federal budget deficit will eventually produce public support for governmental austerity that may adversely impact national security space programs.

References

117th Congress, 1st Session. (2021). *H.R. 5335 To Abolish the Space Force as an Armed Force, and for Other Purposes.* https://www.govinfo.gov/content/pkg/BILLS-117hr5335ih/pdf/BILLS-117hr5335ih.pdf

Barro, R. & De Rugy, V. (2013). *Defense Spending and the Economy.* (Fairfax, VA: Mercatus Center George Mason University). https://www.mercatus.org/system/files/Barro_DefenseSpending_v2.pdf

Biden, J. R. (2021). *Interim National Security Strategic Guidance.* (Washington, D.C.: The White House), 17. https://purl.fdlp.gpo.gov/GPO/LPS152456

Bjørkum, K. (2021). Arctic Strategy: The U.S. and Norwegian Interest and Strategic Effort. *Strategic Studies Quarterly,* 15(3)(Fall): 88–112. https://www.airuniversity.af.edu/Portals/10/SSQ/documents/Volume-15_Issue-3/Bjorkum.pdf.

Congressional Record. (2021(a)). National Defense Authorization Act for Fiscal Year 2022. 167(164) (September 22), H4880-H5076: https://www.govinfo.gov/content/pkg/CREC-2021-09-22/pdf/CREC-2021-09-22-house.pdf

Congressional Record. (2021(b)). National Defense Authorization Act. 167 (165) (September 23): H5103-H5128. https://www.govinfo.gov/content/pkg/CREC-2021-09-23/pdf/CREC-2021-09-23-house.pdf

Garamone, J. (2021). Army, Navy SATCOM Mission Areas Shifting to U.S. Space Force. *DOD News,* (September 22), 1-4: https://www.defense.gov/News/News Stories/Article/Article/2784569/army-navy-satcom-mission-areas-shifting-to-us-space-force/.

Garcia. M. (2021). Supporting United States Space Force. *Congressional Record,* 167 (105) (June 16), H2824: https://www.govinfo.gov/content/pkg/CREC-2021-06-16/pdf/CREC-2021-06-16-house.pdf.

Halaschak, Z. (2021). White House Confirms Space Force Will Continue Under Biden Administration. *Washington Examiner* (February 2, 2021).

Hoehn, J, R., Gallagher, J. C., and Sayler, K. M. (2021). *Overview of Department of Defense Use of the Electromagnetic Spectrum,* Introduction, 10, 17. https://crsreports.congress.gov/product/pdf/R/R46564.

Kirby, L. (2021). USSF, UND sign MOU Establishing University Partnership Program. U.S. Space Force, August 9, 1. https://www.spaceforce.mil/News/Article/2724710/ussf-und-sign-mou-establishing-university-partnership-program/

Listner, M. A. (2021). *Space Weapons: A Briefing with Recommendations for the Biden Administration.* Washington, D.C.: Center for Security Policy, 6-7. https://centerforsecuritypolicy.org/wpcontent/uploads/2021/05/Listner_Space_Weapons_PDF_Optimize-1.pdf

Ludwigson, J. (2021). Space Acquisitions: DOD Faces Challenges and Opportunities Acquiring Space Systems in a Changing Environment. Washington, D.C.: General Accounting Office (May 24): 1-4; 6-17. https://www.gao.gov/assets/gao-

21-520t.pdf

McCall, S. M. (2020). *Defense Primer: National Security Space Launch*. Washington, D.C.: Library of Congress, Congressional Research Service, https://crsreports.congress.gov/product/pdf/IF/IF11531

McCall, S. M. (2021). *Defense Primer: The United States Space Force*. Washington, D.C.: Library of Congress, Congressional Research Service, https://crsreports.congress.gov/product/pdf/IF/IF11495

McNellis, K. P. (2021). *Appropriations Report Language: Overview of Components and Development*. Washington, D.C.: Library of Congress, Congressional Service, A. https://crsreports.congress.gov/product/pdf/R/R44124

Perano, U. (2021) Axios: Democratic Armed Services Chairs Support Space Force After Biden Backing. *Axios* (March 2, 2021), 1-2. https://www.axios.com/space-force-jack-reed-joe-biden-trump-006b35fe-b53c-4856-892d-2270f75b1ebf.html

Poole, C., Bettinger, R., and Reith, M. (2021). Shifting Satellite Control Paradigms: Operational Cybersecurity in the Age of Megaconstellations. *Air and Space Power Journal*, 35(3)(Fall 2021): 46–56.https://www.airuniversity.af.edu/Portals/10/ASPJ/journals/Volume-35_Issue-3/T-Poole.pdf

Public Law 116-92. *National Defense Authorization Act for Fiscal Year 2020*. 133 *U.S. Statutes at Large*, 1561–1563, 2074. https://www.congress.gov/116/plaws/publ92/PLAW-116publ92.pdf

Public Law 116-260, 2021. *H.R. 133 Consolidated Appropriations Act, 2021*. https://www.congress.gov/bill/116th-congress/house-bill/133/text

Public Law 117-103. *H.R. 2471 Consolidated Appropriations Act, 2022*: 109, 120, 122. https://www.congress.gov/bill/117th-congress/house-bill/2471/text

Roth, J. P., Brown Jr., C.Q, and Raymond, J. W. (2021). Department of the Air Force Posture Statement Fiscal Year 2022. (Washington, D.C.: U.S. Congres House Committee on Armed Services: 15–23. https://docs.house.gov/meetings/AS/AS00/20210616/112801/HHRG-117-AS00-Wstate-BrownC-20210616.pdf

Thadalakit, A. (2021). The Effects of Congressional Defense Committees Authorization and Appropriations on Civilian Engineering Employment. Ph.D. Dissertation, University of LaVerne.

Thompson, D. D. (2021). *Fiscal Year 2022 Priorities and Posture of the U.S. Space*

Force, Washington, D.C.: U.S. Congress, House Committee on Armed Services, Subcommittee on Strategic Forces: 2–6. https://docs.house.gov/meetings/AS/AS 29/20210524/112677/HHRG-117-AS29-Wstate-ThompsonD-20210524.pdf

Townsend, B. (2021). The Remote Sensing Revolution Threat." *Strategic Studies Quarterly*, 15 (3)(Fall): 69–87. https://www.airuniversity.af.edu/Portals/10/SSQ/documents/Volume-15_Issue-3/Townsend.pdf.

USAspending.gov. (2021). Space Force. https://www.usaspending.gov/keyword_search/%22space%20force%22

U.S. Congress. House Committee on Appropriations. (2021(b)). *Department of Defense Appropriations Bill, 2022: Together With Minority Views To Accompany H.R. 4432,* House Report 117-88, (Washington, D.C.: GPO)0: 91, 224, 294, 300-303, 49, 353, 357, 359. https://www.govinfo.gov/content/pkg/CRPT-117hrpt88/pdf/CRPT-117hrpt88.pdf

U.S. Congress. House Committee on Appropriations, Defense Subcommittee. (2021(a)). Statement of General Mark Milley, USA, 20th Chairman of the Joint Chiefs of Staff, Department of Defense Budget Hearing. (May 27): 8. https://docs.house.gov/meetings/AP/AP02/20210527/112682/HHRG-117-AP02-Wstate-MilleyM-20210527.pdf

U.S. Congress. House Committee on Armed Services. (2021). *National Defense Authorization Act For Fiscal Year, 2022, H.R. 4350 Together With Additional and Dissenting Views,* House Report 117-118, (Washington, D.C.: GPO): September 10: 61–62, 124, 134, 153, 274, 276–279, 292, 346, 391–392, 433, 436, 469, 570. https://www.govinfo.gov/content/pkg/CRPT-117hrpt118/pdf/CRPT-117hrpt118.pdf

U.S. Congress, Senate Committee on Appropriations, Subcommittee on Defense. (2021). Statement by Ms. Barbara McQuiston Performing the Duties of the Under Secretary of Defense for Research and Engineering: 22, 25. https://www.appropriations.senate.gov/imo/media/doc/McQuiston%20Statement%20for%20the%20Record.pdf

U.S. Congress, Senate Committee on Armed Services. (2021(d)). To Receive Testimony on United States Strategic Command and United States Space Command in Review of the Defense Authorization Request For Fiscal Year 2022 and the Future Years Defense Program. Washington, D.C.: Senate Armed Services Committee, 27–29, 38–39, 43–44, 50–53, 83–84, 88–89. https://www.armed-services.senate.gov/imo/media/doc/21-22_04-20-2021.pdf

U.S. Congress, Senate Armed Services Committee. (2021(c)). Advance Policy Questions for Dr. Colin Kahl Nominee for Appointment to be Under Secretary of Defense for Policy. Washington, D.C.: Senate Armed Services Committee, https://www.armed-services.senate.gov/imo/media/doc/Kahl_APQs_03-04-21.pdf; 12–15.

U.S. Congress, Senate Armed Services Committee. (2021(a)). Advance Policy Questions for Lloyd J. Austin Nominee for Appointment to be Secretary of Defense: 54–57; 122. https://www.armed-services.senate.gov/imo/media/doc/Austin_APQs_01-19-21.pdf

U.S. Congress, Senate Armed Services Committee. (2021(b)). Advance Policy Questions for Dr. Kathleen Hicks Nominee for Appointment to be Deputy Secretary of Defense, Washington, D.C.: Senate Armed Services Committee: 70–76. https://www.armed-services.senate.gov/imo/media/doc/Hicks_APQs_02-02-21.pdf

U.S. Congress, Senate Armed Services Committee. (2021(e)). *National Defense Authorization Act for Fiscal Year 2022*, Senate Report 117-39. Washington, D.C.: GPO: 73, 145, 296–300, 378–379, 522. https://www.govinfo.gov/content/pkg/CRPT-117srpt39/pdf/CRPT-117srpt39.pdf;

U.S. Congress, Senate Armed Services Committee. (2021). *To Receive Testimony on Space Force, Military Space Operations and Programs*. (May 26): 2, 5–9, 11–12, 15–21, 23, 25–47, 51, 53–55, 57–59. https://www.armed-services.senate.gov/imo/media/doc/21-44_05-26-2021.pdf

U.S. Defense Intelligence Agency. (2022). *2022 Challenges to Security inSpace: Space Reliance in an Era of Competition and Expansion*. Washington, D.C.: Defense Intelligence Agency: iii–iv, 8–11, 20–28, 30–32. https://www.dia.mil/Portals/110/Documents/News/Military_Power_Publications/Challenges_Security_Space_2022.pdf

U.S. Department of Defense, Director, Operational Test and Evaluation. (2021). *FY 2020 Annual Report*, Washington, D.C.: Director, Operational Test and Evaluation: III, 191–194, 205–206, 238–239. https://www.dote.osd.mil/Publications/Annual-Reports/2020-Annual-Report/

U.S. Federal Contractor (USCFR). (2021(a)). *STPSTEP 2.0 Program*: 1. https://usfcr.com/search/opportunities/?oppId=5816666c540c4e46b2dab42f2cccf327

USCFR. (2021(b)). *USSF SWAC Business Fair*: 1. https://usfcr.com/search/opportunities/?oppId=1336b81ec0ce49909cd43fbb533856e6

USCFR. (2021(c)). *Peterson Space Force Ballistic Glass Panel Requirement*: 1. https://usfcr.com/search/opportunities/?oppId=3edebda2fc1a48c4945cc81cd1891546

U.S. Government Accountability Office. (2021). *Decision Matter of Inmarsat, Government, Inc.* Washington, D.C.: GAO, https://www.gao.gov/assets/b-419583.pdf

U.S. Department of Defense, Office of the Under Secretary of Defense (Comptroller). (2021). *National Defense Budget Estimates for FY 2022* (August), 46–48, 51, 76--77, 110–111, 117–118, 260.

U.S. Office of Management and Budget. (2021). *Appendix: Budget of the United States Government: Fiscal Year 2022.* (Washington, D.C.: Office of Management and Budget, 2021): 248, 280, 289. https://www.govinfo.gov/content/pkg/BUDGET-2022-APP/pdf/BUDGET-2022-APP.pdf

U.S. Space Force. (2021(a)). *U.S. Space Force Vision for a Digital Service.* (May): 2, 3–7, 9–13. https://purl.fdlp.gov/GPO/gpo159289

U.S. Space Force. (2021(b)). "USSF, UND sign MOU Establishing University Partnership Program." *Space Force News* (August 9): 1–2. https://www.spaceforce.mil/News/Article/2724710/ussf-und-sign-mou-establishing-university-partnership-program/

Weires, S. G. (2021). Fla. Recruiter Helps Build Space Force Military Branch. *The Palm Beach Post* (August 15, 2021): B9.

Wolfgang, B. (2020). Joe Biden Pressured to Scrap Space Force. *Washington Times*, (November 18, 2020): 1-6; https://www.washingtontimes.com/news/2020/nov/18/joe-biden-pressured-scrap-space-force/

Related Titles from Westphalia Press

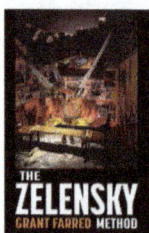

The Zelensky Method
by Grant Farred

Locating Russian's war within a global context, The Zelensky Method is unsparing in its critique of those nations, who have refused to condemn Russia's invasion and are doing everything they can to prevent economic sanctions from being imposed on the Kremlin.

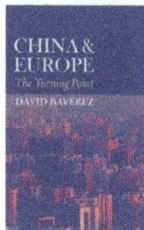

China & Europe: The Turning Point
by David Baverez

In creating five fictitious conversations between Xi Jinping and five European experts, David Baverez, who lives and works in Hong Kong, offers up a totally new vision of the relationship between China and Europe.

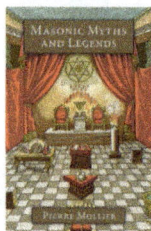

Masonic Myths and Legends
by Pierre Mollier

Freemasonry is one of the few organizations whose teaching method is still based on symbols. It presents these symbols by inserting them into legends that are told to its members in initiation ceremonies. But its history itself has also given rise to a whole mythology.

Resistance: Reflections on Survival, Hope and Love
Poetry by William Morris, Photography by Jackie Malden

Resistance is a book of poems with photographs or a book of photographs with poems depending on your perspective. The book is comprised of three sections titled respectively: On Survival, On Hope, and On Love.

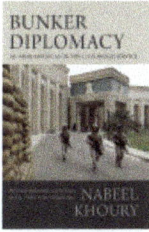

Bunker Diplomacy: An Arab-American in the U.S. Foreign Service
by Nabeel Khoury

After twenty-five years in the Foreign Service, Dr. Nabeel A. Khoury retired from the U.S. Department of State in 2013 with the rank of Minister Counselor. In his last overseas posting, Khoury served as deputy chief of mission at the U.S. embassy in Yemen (2004-2007).

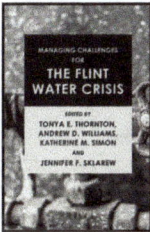

Managing Challenges for the Flint Water Crisis
Edited by Toyna E. Thornton, Andrew D. Williams, Katherine M. Simon, Jennifer F. Sklarew

This edited volume examines several public management and intergovernmental failures, with particular attention on social, political, and financial impacts. Understanding disaster meaning, even causality, is essential to the problem-solving process.

Donald J. Trump, The 45th U.S. Presidency and Beyond International Perspectives
Editors: John Dixon and Max J. Skidmore

The reality is that throughout Trump's presidency, there was a clearly perceptible decline of his—and America's—global standing, which accelerated as an upshot of his mishandling of both the Corvid-19 pandemic and his 2020 presidential election loss.

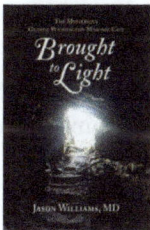

Brought to Light: The Mysterious George Washington Masonic Cave
by Jason Williams, MD

The George Washington Masonic Cave near Charles Town, West Virginia, contains a signature carving of George Washington dated 1748. Although this inscription appears authentic, it has yet to be verified by historical accounts or scientific inquiry.

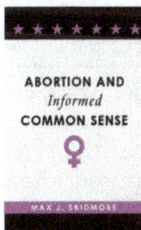

Abortion and Informed Common Sense
by Max J. Skidmore

The controversy over a woman's "right to choose," as opposed to the numerous "rights" that abortion opponents decide should be assumed to exist for "unborn children," has always struck me as incomplete. Two missing elements of the argument seems obvious, yet they remain almost completely overlooked.

The Athenian Year Primer: Attic Time-Reckoning and the Julian Calendar
by Christopher Planeaux

The ability to translate ancient Athenian calendar references into precise Julian-Gregorian dates will not only assist Ancient Historians and Classicists to date numerous historical events with much greater accuracy but also aid epigraphists in the restorations of numerous Attic inscriptions.

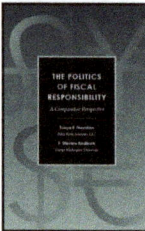

The Politics of Fiscal Responsibility: A Comparative Perspective
by Tonya E. Thornton and F. Stevens Redburn

Fiscal policy challenges following the Great Recession forced members of the Organisation for Economic Co-operation and Development (OECD) to implement a set of economic policies to manage public debt.

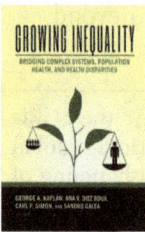

Growing Inequality: Bridging Complex Systems, Population Health, and Health Disparities
Editors: George A. Kaplan, Ana V. Diez Roux, Carl P. Simon, and Sandro Galea

Why is America's health is poorer than the health of other wealthy countries and why health inequities persist despite our efforts? In this book, researchers report on groundbreaking insights to simulate how these determinants come together to produce levels of population health and disparities and test new solutions.

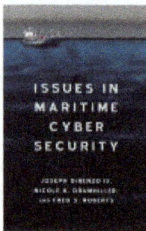

Issues in Maritime Cyber Security
Edited by Dr. Joe DiRenzo III, Dr. Nicole K. Drumhiller, and Dr. Fred S. Roberts

The complexity of making MTS safe from cyber attack is daunting and the need for all stakeholders in both government (at all levels) and private industry to be involved in cyber security is more significant than ever as the use of the MTS continues to grow.

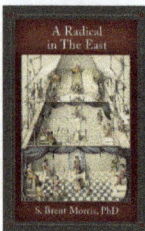

A Radical In The East
by S. Brent Morris, PhD

The papers presented here represent over twenty-five years of publications by S. Brent Morris. They explore his many questions about Freemasonry, usually dealing with origins of the Craft. A complex organization with a lengthy pedigree like Freemasonry has many basic foundational questions waiting to be answered, and that's what this book does: answers questions.

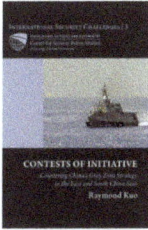

Contests of Initiative: Countering China's Gray Zone Strategy in the East and South China Seas
by Dr. Raymond Kuo

China is engaged in a widespread assertion of sovereignty in the South and East China Seas. It employs a "gray zone" strategy: using coercive but sub-conventional military power to drive off challengers and prevent escalation, while simultaneously seizing territory and asserting maritime control.

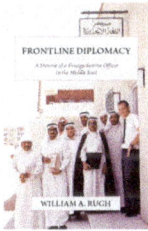

Frontline Diplomacy: A Memoir of a Foreign Service Officer in the Middle East
by William A. Rugh

In short vignettes, this book describes how American diplomats working in the Middle East dealt with a variety of challenges over the last decades of the 20th century. Each of the vignettes concludes with an insight about diplomatic practice derived from the experience.

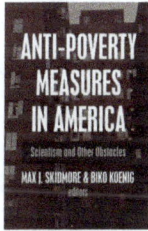

Anti-Poverty Measures in America: Scientism and Other Obstacles
Editors, Max J. Skidmore and Biko Koenig

Anti-Poverty Measures in America brings together a remarkable collection of essays dealing with the inhibiting effects of scientism, an over-dependence on scientific methodology that is prevalent in the social sciences, and other obstacles to anti-poverty legislation.

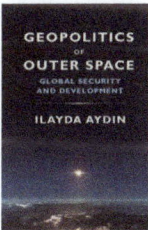

Geopolitics of Outer Space: Global Security and Development
by Ilayda Aydin

A desire for increased security and rapid development is driving nation-states to engage in an intensifying competition for the unique assets of space. This book analyses the Chinese-American space discourse from the lenses of international relations theory, history and political psychology to explore these questions.

westphaliapress.org

Policy Studies Organization

www.ingramcontent.com/pod-product-compliance
Lightning Source LLC
LaVergne TN
LVHW061328060426
835511LV00012B/1913